NATIONAL ACADEMIES *Sciences Engineering Medicine*

NATIONAL ACADEMIES PRESS
Washington, DC

Developing a Research Agenda on Contrails and Their Climate Impacts

Committee on the Research Agenda for Reducing the Climate Impact of Aviation-Induced Cloudiness and Persistent Contrails from Commercial Aviation

Aeronautics and Space Engineering Board

Space Studies Board

Division on Engineering and Physical Sciences

Consensus Study Report

NATIONAL ACADEMIES PRESS 500 Fifth Street, NW Washington, DC 20001

This activity was supported by a contract between the National Academy of Sciences and the National Aeronautics and Space Administration. Any opinions, findings, conclusions, or recommendations expressed in this publication do not necessarily reflect the views of any organization or agency that provided support for the project.

International Standard Book Number-13: 978-0-309-73551-3
Digital Object Identifier: https://doi.org/10.17226/29073

This publication is available from the National Academies Press, 500 Fifth Street, NW, Keck 360, Washington, DC 20001; (800) 624-6242; http://nap.nationalacademies.org.

The manufacturer's authorized representative in the European Union for product safety is Authorised Rep Compliance Ltd., Ground Floor, 71 Lower Baggot Street, Dublin D02 P593 Ireland; www.arccompliance.com.

Copyright 2025 by the National Academy of Sciences. National Academies of Sciences, Engineering, and Medicine and National Academies Press and the graphical logos for each are all trademarks of the National Academy of Sciences. All rights reserved.

Printed in the United States of America.

Suggested citation: National Academies of Sciences, Engineering, and Medicine. 2025. *Developing a Research Agenda on Contrails and Their Climate Impacts*. Washington, DC: National Academies Press. https://doi.org/10.17226/29073.

The **National Academy of Sciences** was established in 1863 by an Act of Congress, signed by President Lincoln, as a private, nongovernmental institution to advise the nation on issues related to science and technology. Members are elected by their peers for outstanding contributions to research. Dr. Marcia McNutt is president.

The **National Academy of Engineering** was established in 1964 under the charter of the National Academy of Sciences to bring the practices of engineering to advising the nation. Members are elected by their peers for extraordinary contributions to engineering. Dr. John L. Anderson is president.

The **National Academy of Medicine** (formerly the Institute of Medicine) was established in 1970 under the charter of the National Academy of Sciences to advise the nation on medical and health issues. Members are elected by their peers for distinguished contributions to medicine and health. Dr. Victor J. Dzau is president.

The three Academies work together as the **National Academies of Sciences, Engineering, and Medicine** to provide independent, objective analysis and advice to the nation and conduct other activities to solve complex problems and inform public policy decisions. The National Academies also encourage education and research, recognize outstanding contributions to knowledge, and increase public understanding in matters of science, engineering, and medicine.

Learn more about the National Academies of Sciences, Engineering, and Medicine at **www.nationalacademies.org**.

Consensus Study Reports published by the National Academies of Sciences, Engineering, and Medicine document the evidence-based consensus on the study's statement of task by an authoring committee of experts. Reports typically include findings, conclusions, and recommendations based on information gathered by the committee and the committee's deliberations. Each report has been subjected to a rigorous and independent peer-review process and it represents the position of the National Academies on the statement of task.

Proceedings published by the National Academies of Sciences, Engineering, and Medicine chronicle the presentations and discussions at a workshop, symposium, or other event convened by the National Academies. The statements and opinions contained in proceedings are those of the participants and are not endorsed by other participants, the planning committee, or the National Academies.

Rapid Expert Consultations published by the National Academies of Sciences, Engineering, and Medicine are authored by subject-matter experts on narrowly focused topics that can be supported by a body of evidence. The discussions contained in rapid expert consultations are considered those of the authors and do not contain policy recommendations. Rapid expert consultations are reviewed by the institution before release.

For information about other products and activities of the National Academies, please visit www.nationalacademies.org/about/whatwedo.

COMMITTEE ON THE RESEARCH AGENDA FOR REDUCING THE CLIMATE IMPACT OF AVIATION-INDUCED CLOUDINESS AND PERSISTENT CONTRAILS FROM COMMERCIAL AVIATION

TIMOTHY C. LIEUWEN (NAE), Georgia Institute of Technology, *Chair*
STEVE BARRETT, Cambridge University
SEAN BRADSHAW, Pratt & Whitney
LETICIA CUELLAR-HENGARTNER, Los Alamos National Laboratory
ERIC H. DUCHARME (NAE), Martlet Engineering, LLC
ANDREW GETTELMAN, Pacific Northwest National Laboratory
ROBERT J. HANSMAN, JR. (NAE), Massachusetts Institute of Technology
RICHARD H. MOORE, NASA Langley Research Center
JOYCE E. PENNER, University of Michigan
MICHAEL J. PRATHER, University of California, Irvine

Staff

DWAYNE A. DAY, Senior Program Officer, Aeronautics and Space Engineering Board
LINDA M. WALKER, Program Coordinator, Aeronautics and Space Engineering Board, Board on Physics and Astronomy, and Space Studies Board
DIONNA WISE, Program Coordinator, Space Studies Board

COLLEEN N. HARTMAN, Senior Board Director, Space Studies Board, Aeronautics and Space Engineering Board, and Board on Physics and Astronomy (through May 15, 2025)

AERONAUTICS AND SPACE ENGINEERING BOARD

ILAN KROO (NAE), Stanford University, *Chair*
SEAN BRADSHAW, Pratt & Whitney
ROBERT D. BRAUN (NAE), Johns Hopkins University Applied Physics Laboratory
EDWARD F. CRAWLEY (NAE), Massachusetts Institute of Technology
SHANA L. DALE, National Aeronautics and Space Administration
ERIC H. DUCHARME (NAE), Martlet Engineering, LLC
WILLIAM R. GRAY, U.S. Air Force
MORIBA K. JAH, University of Texas at Austin
JOHN C. KARAS, Lockheed Martin Space Systems Company
TIMOTHY C. LIEUWEN (NAE), Georgia Institute of Technology
GEORGE T. LIGLER (NAE), Texas A&M University
GENERAL LESTER L. LYLES (NAE), U.S. Air Force (retired)
ELLEN OCHOA (NAE), NASA Johnson Space Center
TOM G. REYNOLDS, MIT Lincoln Laboratory
WANDA A. SIGUR (NAE), Lockheed Martin Corporation
MICHAEL K. SINNETT (NAE), Boeing Commercial Airplanes
ANTHONY M. WAAS, Arizona State University

Staff

COLLEEN N. HARTMAN, Senior Board Director, Space Studies Board, Aeronautics and Space Engineering Board, and Board on Physics and Astronomy (through May 15, 2025)
ARUL MOZHI, Associate Board Director (Acting Board Director from May 16, 2025)
TANJA PILZAK, Manager, Program Operations
ANDREA REBHOLZ, Program Coordinator

SPACE STUDIES BOARD

MARGARET G. KIVELSON (NAS), University of California, Los Angeles, *Chair*
JAMES H. CROCKER (NAE), Lockheed Martin Corporation, *Vice Chair*
DANIELA CALZETTI (NAS), University of Massachusetts Amherst
ROBIN M. CANUP (NAS), Southwest Research Institute
DEEPTO CHAKRABARTY, Massachusetts Institute of Technology
JEFF DOZIER, University of California, Santa Barbara
MELINDA D. DYAR, Mount Holyoke College
ANTONIO L. ELIAS (NAE), Orbital ATK Inc.
STEPHEN J. MACKWELL, National Science Foundation
PETER I. MESZAROS (NAS), Pennsylvania State University, University Park
RICHARD M. OBERMANN, National Research Council
NELSON PEDREIRO (NAE), Lockheed Martin Space Systems Company
CHRISTA D. PETERS-LIDARD (NAE), NASA Goddard Space Flight Center
MARK P. SAUNDERS, Independent Consultant
HOWARD J. SINGER, National Oceanic and Atmospheric Administration
KEIVAN G. STASSUN, Vanderbilt University
ERIKA B. WAGNER, Blue Origin, LLC
PAUL D. WOOSTER, SpaceX
ENDAWOKE YIZENGAW, The Aerospace Corporation
GARY P. ZANK (NAS), University of Alabama in Huntsville

Staff

COLLEEN N. HARTMAN, Senior Board Director, Space Studies Board, Aeronautics and Space Engineering Board, and Board on Physics and Astronomy (through May 15, 2025)
ARUL MOZHI, Associate Board Director (Acting Board Director from May 16, 2025)
TANJA PILZAK, Manager, Program Operations
ANDREA REBHOLZ, Program Coordinator

Reviewers

This Consensus Study Report was reviewed in draft form by individuals chosen for their diverse perspectives and technical expertise. The purpose of this independent review is to provide candid and critical comments that will assist the National Academies of Sciences, Engineering, and Medicine in making each published report as sound as possible and to ensure that it meets the institutional standards for quality, objectivity, evidence, and responsiveness to the study charge. The review comments and draft manuscript remain confidential to protect the integrity of the deliberative process.

We thank the following individuals for their review of this report:

IRENE C. DEDOUSSI, University of Cambridge
BRANDON GRAVER, Airlines for America
JIM HILEMAN, Boeing
GEORGE T. LIGLER (NAE), Texas A&M University
RICK MIAKE-LYE, Aerodyne Research Inc.
ANDREW ROLLINS, National Oceanic and Atmospheric Administration
MARC SHAPIRO, Breakthrough Energy
LAURENCE VIGEANT-LANGLOIS, AE Industrial Partners
CHRISTIANE VOIGT, German Aerospace Center (DLR)
DONALD WUEBBLES, University of Illinois Urbana-Champaign

Although the reviewers listed above provided many constructive comments and suggestions, they were not asked to endorse the conclusions or recommendations of this report nor did they see the final draft before its release. The review of this report was overseen by **BRADLEY COLMAN**, Bayer Corporation (retired), and **CHRIS FIELD (NAS)**, Stanford University. They were responsible for making certain that an independent examination of this report was carried out in accordance with the standards of the National Academies and that all review comments were carefully considered. Responsibility for the final content rests entirely with the authoring committee and the National Academies.

Contents

PREFACE   xiii

SUMMARY   1

1   OVERVIEW   9
      Report Outline, 16
      References, 18

2   AIRCRAFT ENGINE EMISSIONS   19
      General Combustor Emissions, 19
      Lubrication Oil Vent Effects, 28
      Particle Impacts on Contrail Formation and Downstream Aerosol-Cloud Interactions, 29
      References, 32

3   ATMOSPHERIC MEASUREMENTS   34
      Measurements of Atmospheric State Parameters, 34
      Measurements of Contrails and Contrail Cirrus, 48
      Measurements of Atmospheric Particles, 50
      References, 53

4   CONTRAIL MODELING SYSTEMS   56
      Wake Vortex Models, 56
      Global Climate Models, 57
      Contrail Plume Models, 57
      Ice-Supersaturation Forecast Models, 58
      System for Contrail Prediction, 58
      Summary, 58
      References and Further Reading, 59

5 CONTRAIL FORECAST AND VERIFICATION   61
   The Purpose of a Contrail Forecast System, 61
   Existing Contrail Forecast Systems, 63
   Ice-Supersaturated Region Nowcasting/Forecasting, 63
   Contrail Forecasting, 64
   Evaluation Methods, 65
   What Would a Forecast System Look Like?, 66
   Challenges and Opportunities, 68
   References, 70

6 OPERATIONAL CONCEPTS   71
   References and Further Reading, 75

APPENDIXES

A Statement of Task   77
B Committee and Staff Biographical Information   79
C Recommendations   83

Preface

In early 2024, NASA requested that the National Academies of Sciences, Engineering, and Medicine conduct a study to develop a national research agenda to better understand, quantify, and support the development of technical and operational solutions to significantly reduce the global climate impact of aviation-induced cloudiness and persistent contrails from commercial aviation. Under the auspices of the Aeronautics and Space Engineering Board and the Space Studies Board, the National Academies created the Committee on the Research Agenda for Reducing the Climate Impact of Aviation-Induced Cloudiness and Persistent Contrails from Commercial Aviation to undertake this task. The committee began meeting in summer 2024, holding multiple virtual meetings to gather data from a wide range of experts in government, industry, and academia, as well as international perspectives. The committee heard from representatives of airlines and aircraft manufacturers as well as atmospheric scientists and propulsion engineers. By fall 2024, the committee began writing its report and held in-person and virtual meetings to generate findings and recommendations, guided by its statement of task (see Appendix A).

Summary

Aviation has significant global climate impacts that stem from the interaction of jet engine exhaust with the atmosphere. Aviation CO_2 emissions from burning fossil jet fuel are about 2.5 percent of total anthropogenic CO_2 emissions and are a contributor to total anthropogenic climate forcing. Aviation non-CO_2 climate effects are estimated to be the same order of magnitude as aviation CO_2 climate effects. One of the largest non-CO_2 climate effects of aviation is due to persistent condensation trails (contrails) and aviation-induced cirrus clouds. Contrails form when the cooling aircraft engine exhaust plume becomes supersaturated with respect to liquid water (i.e., relative humidity exceeds 100 percent), causing the formation of many small water droplets that freeze to create a visible white trail. When the surrounding atmosphere is supersaturated with respect to ice, these initial contrails can grow and persist into contrail cirrus clouds for hours (Figure S-1); otherwise, the contrail sublimates. Persistent contrail cirrus, like naturally occurring cirrus clouds, will both scatter incoming sunlight back to space (cooling) and trap Earth's outgoing thermal radiation (warming). Nighttime contrails are always warming, while daytime contrails can be either cooling or warming depending on the atmospheric and surface conditions as well as the optical properties of the contrail. Modeling both of these radiative effects shows that overall, considering all the persistent contrails generated by the current aircraft fleet, persistent contrails and aviation-induced cirrus create a warming effect. Furthermore, since only persistent contrail cirrus are climatically relevant, it is critical to be able to forecast and diagnose these ice-supersaturated regions (ISSRs) of the atmosphere, which are thought to be horizontally vast (~hundreds of kilometers) but vertically shallow (~hundreds of meters). Throughout this report, the terms contrails, persistent contrails, contrail cirrus, and aviation-induced cloudiness are used synonymously.

The United States has been actively engaged and leading in all aspects of aviation, including contributing to cutting-edge research on the science of flight and its environmental effects. The U.S. government, in coordination with industry and academia, has tremendous resources to assist industry in making intelligent choices to mitigate the environmental impact of aviation, and maintain and enhance global competitiveness. In recent years, international organizations have begun to regulate various aspects of these impacts, affecting both U.S. commercial airlines and aircraft manufacturers. The United States needs sufficient data on the causes and effects of persistent contrails, and possible mitigating actions, to enable a response to this emerging regulatory environment. There is strong industry support for emissions and contrails research and an opportunity to enhance U.S. market differentiation and economic competitiveness.

In early 2024, NASA tasked the National Academies of Sciences, Engineering, and Medicine to develop a national research agenda that enhances the scientific understanding of persistent contrails and aviation-induced

FIGURE S-1 The United States has significant and unique capabilities to be a leader in contrails research. The four-engine NASA DC-8 flying laboratory is shown here generating contrails while sampling the contrails from a two-engine Boeing ecoDemonstrator Explorer 737-10. The flight test campaign took place during October 2023.
SOURCE: Copyright © Boeing.

cloudiness and accelerates the development of technical and operational solutions for reducing their climate impacts. NASA has been the lead U.S. government agency for over a decade in the study of persistent contrails and aviation-induced cloudiness and has been working in close collaboration with researchers across the U.S. government and around the globe. It has flown several research campaigns to observe the generation of contrails, including as recently as November 2024. In addition, NASA has a long history of the development of Earth-observing spacecraft as well as data products that have been used by operational agencies, such as the National Oceanic and Atmospheric Administration (NOAA), and has developed sensors that can be used by commercial aircraft to measure atmospheric conditions related to contrail formation. NASA has also been involved in the development of new propulsion technologies and aviation fuels and has close ties with industry in these areas. Finally, NASA has a long-standing relationship with the Federal Aviation Administration (FAA), including the development of air traffic control (ATC) technologies, experience, and connections that will be valuable in future contrail mitigation strategies.

The Committee on the Research Agenda for Reducing the Climate Impact of Aviation-Induced Cloudiness and Persistent Contrails from Commercial Aviation was established in late spring 2024 and soon began data-gathering sessions and meetings with commercial industry, U.S. government agencies, regulators, and international organizations throughout the second half of 2024 to develop a specific set of prioritized research recommendations. The committee's statement of task is included in Appendix A.

The bulk of the warming effect of contrails results from a relatively small percentage of all flights. Further research will enable better understanding of the causes of persistent contrails, methods to model their effects, and emerging factors like the development of new propulsion systems and synthetic fuels that affect the generation of contrails. International organizations have already begun the process of measuring, reporting, and verifying the generation of emissions including contrails. Without significant research, the United States is at risk of becoming less competitive with its commercial aviation industry if, for instance, it does not develop improved technologies and operational procedures to respond to emerging regulations.

There are several ways to characterize the location and extent of contrails, including space-based sensors and in situ sensors that can be mounted on commercial aircraft fleets. The primary uncertainty in the formation and persistence of contrails is the ability to observe and/or predict ISSRs where the relative humidity over ice (RHi) is greater than 100 percent. RHi along flight paths is calculated from upper tropospheric temperature and humidity. Accurate, high-resolution vertical and horizontal measurements of humidity are needed to constrain model forecasts and nowcasts of cruise-level ISSRs and contrail-forming conditions.

Atmospheric modeling systems already exist and are used for multiple purposes such as weather prediction, climate prediction, and air quality. Individual contrails are often modeled using very detailed and small-scale plume models. Only a few atmospheric modeling systems include contrails in them. As contrail forecasting gains in importance and more data about them are collected, models of the atmosphere and contrails need to be linked into a seamless system so that differing models and meteorological data can be tested and utilized for contrail prediction.

The major gap in our understanding of the location and occurrence of persistent contrails is the lack of high-precision humidity observations in aircraft flight regions of the upper troposphere (temperature is already measured from aircraft). Accuracy of 1–2 parts per million (ppm) is desired down to a lower detection limit of 20–30 ppm. Current in situ sensor systems for commercial aircraft lack the reliability and calibration stability necessary for widespread deployment across the commercial fleet. Sensors need to integrate with existing aircraft data downlink systems and be certified for new and existing aircraft. Deployment of a large number of water vapor sensors across the commercial airline fleet would be invaluable to simulate and forecast cruise-level ice supersaturation for NOAA, FAA, and other national and international stakeholders.

As previously noted, there is confidence that persistent contrails create a warming effect. Even though there is ±70 percent uncertainty in the magnitude of climate warming from contrails, the sign of the warming effect (i.e., warming versus cooling) is robust when integrated across the whole fleet. Uncertainty in the climate effect of persistent contrails for any single flight, however, remains large. Contrail effects with a short lifetime (hours) need to be compared to effects of CO_2 emissions with a long lifetime (decades to centuries), and the metrics for comparison have both physical and social assumptions (timescale of interest).

Better data on contrail formation can enable mitigation strategies. There are several methods for reducing or mitigating the climate impact of persistent contrails. Changing how aircraft operate to avoid the generation of high-impact contrails is one possible method. Some specific contrails contribute notably to climate warming, and mitigating them through operational measures in situations that do not introduce significant additional CO_2 emissions is a robust strategy regardless of the selected metric. Operational contrail mitigation requires working within the current air traffic control system to safely reroute aircraft to avoid contrail formation regions. Elements of the current air traffic control system could limit the ability of aircraft to execute effective avoidance strategies while maintaining safety. Other potential mitigation factors could be through the use of alternate fuels and advanced engine combustion technologies that could reduce contrail formation and persistence.

The recommendations in this report represent the priorities for a national contrails research strategy and provide a vision for how research could eventually support operational contrails mitigation. The committee has further prioritized them into key short-term and long-term research priorities, and other priorities.

KEY RESEARCH PRIORITIES

The highest-priority research areas are listed in this section. While all these items are high priority and should ideally be initiated as soon as possible, they are organized into items that will have near-term impacts and long-term impacts, as well as places where NASA can make a real and unique difference. Note that long-term impacts may start now, but will continue and require more integration with others.

Near-Term Impacts

There are several key actions—using current observations, forecasting models, and the aviation system—that could be done in the near term to advance the understanding of contrails and demonstrate effective contrail mitigation strategies.

The primary uncertainty in forecasting contrails is predicting the formation and evolution of persistent contrails, which occur in ISSRs of the atmosphere, which requires upper tropospheric temperature and humidity measurements. Existing forecast systems can partially predict ISSRs, but enhancements to vertical resolution and more accurate RHi forecasts are critical. To accurately predict ice supersaturation, better observations of humidity at flight altitudes are needed. This will likely require improving humidity sensors for commercial aircraft, deploying them, and using this information in models that predict where contrails are likely to form. Mapping observations of persistent contrails with aircraft location data can also improve the understanding of ISSRs.

Short-Term Priority Recommendations

Recommendation: NASA should support the development, testing, and certification of advanced and accurate commercial-aircraft-capable humidity and temperature sensors for contrail-forming regions as well as onboard contrail-detecting cameras and automated contrail-detection image-recognition algorithms. (Chapter 3)

Recommendation: NASA should support research and observational studies to improve the understanding of the extent and frequency of ice-supersaturated regions (ISSRs) and the level of skill in simulating ISSRs and contrails. (Chapter 3)

Recommendation: NASA should apply its current Earth system modeling efforts in support of simulating ice-supersaturated regions and contrails as a pathway to demonstrate the use of observations and advanced modeling tools for developing a contrail forecast and prediction system and estimating contrail radiative forcing. (Chapter 5)

Persistent contrails are not a new phenomenon, and were commonly observed during World War II, but the scientific study of them has increased in recent decades (Figure S-2). As indicated in the chapters of this report, NASA is already engaged in some of these activities, although there are also transitions under way (e.g., NASA recently retired its DC-8 research aircraft and its replacement will not be available until 2026).

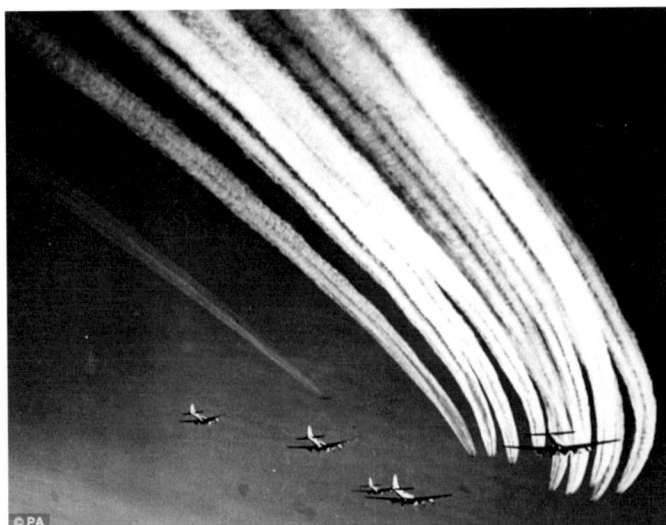

FIGURE S-2 Contrails, and their persistence, are not a new phenomenon. They were first observed in 1940, and during World War II high-altitude bombing raids like this one often generated contrails.
SOURCE: Courtesy of the U.S. Air Force Historical Research Agency. Photo by Unknown.

Long-Term Impacts

There are several priorities for starting research now to benefit long-term strategies to constrain and reduce the radiative forcing of persistent contrails and aviation-induced cloudiness. Research efforts to evaluate the impact of sustainable aviation fuel chemistry and low-emission engine technologies on the particulate emissions that influence contrail formation, evolution, and radiative forcing should be prioritized. Moreover, observations of aviation emissions and background particulates (ambient aerosols) at flight altitudes are critical to understand the background atmosphere and clouds at flight levels, supporting the long-term deployment of advanced aircraft engines and sustainable aviation fuels that reduce contrail radiative forcing. Lastly, ensuring the development of large-scale systems for observing contrails is also important for the validation of physics-based models and for constraining estimates of contrail radiative forcing.

Long-Term Priority Recommendations

Recommendation: NASA, in coordination with the Federal Aviation Administration, the Department of Energy, and the Department of Defense, should support laboratory and engine research studies to improve the understanding of how fuel composition, combustor technology, and engine operating conditions impact particulate emissions (volatile and non-volatile) and contrail properties. (Chapter 2)

Recommendation: NASA should continue to collect in-flight observational data of contrails, cruise emissions (CO_2, NOx, and ice-nucleating particles) from aviation that advance the understanding of the factors that influence contrail properties. (Chapter 2)

Recommendation: NASA should identify and enable a minimum set of key aerosol instruments that can be flown on multiple missions with the goal of characterizing the aerosol composition of the upper troposphere and uncovering the contribution of aviation emissions relative to other sources. (Chapter 3)

AIRCRAFT ENGINE EMISSIONS

Contrail ice crystals form on particles emitted from the aircraft engines, as well as ambient particles in the upper troposphere. Aircraft engine particulate emissions influence contrail properties and resultant radiative forcing. Aircraft engine particulate emissions are strongly influenced by fuel composition, engine operating conditions, lubrication oil venting, combustion system technology, and maintenance cycle. Because of the importance of particulate emissions in contrail dynamics, advanced engine technologies that reduce particulate emissions can play a key role in mitigating contrail radiative forcing. These technology levers include advanced combustor designs, vent oil management, and alternative fuels.

The largest impact of alternative fuels on contrails is likely to be through changes in the particulate content of aircraft exhaust. Developing the ability to identify the composition and roles of all these contributors is needed to predict how changes in fuel composition and engine technology will influence engine particle emissions relevant for contrail formation.

Recommendation: NASA, in coordination with the Federal Aviation Administration, the Department of Energy, the Department of Defense, other relevant federal agencies, and the private sector, should support development of low-particle-emitting combustion technologies, as well as sustainable aviation fuels with inherently low particulate-formation tendencies. (Chapter 2)

ATMOSPHERIC MEASUREMENTS

The number and distribution of necessary sensors deployed across the fleet have yet to be optimized and will depend on the operational avoidance and verification goals and whether these are to be realized at the individual flight or fleet level.

Recommendation: NASA should support observing system simulation experiments to define widespread water vapor sensor deployment to best inform contrail forecast systems and individual verification and avoidance efforts. (Chapter 3)

NASA is uniquely positioned to test novel in situ temperature and humidity sensors using existing Science Mission Directorate Airborne Science Program aircraft and state-of-the-art research instruments.

Satellites also can provide necessary data for contrail prediction and modeling. Next-generation geostationary satellite sounders that may be relevant for measuring flight-level temperature and humidity and tracking persistent contrails will launch over the coming decade. Low Earth orbit satellites and constellations also show strong promise for inferring temperature, humidity, and persistent contrail tracking. NASA fills an important research role within the United States in developing and demonstrating satellite data products and assimilating these products into models that can eventually be deployed to operational agencies (e.g., NOAA, FAA).

Satellite- and ground-based imagers and airborne lidars, automated detection algorithms, and flight trajectory information are important tools for validating contrail model predictions. Artificial intelligence and machine learning are becoming more widely used for automated contrail observation and detection from satellite and ground-based imagery. This research is ongoing and is necessary for informing satellite design decisions.

Recommendation: NASA should support satellite remote sensing research for diagnosing persistent contrails and ice-supersaturated regions to develop readiness for the next-generation geostationary sounders and imagers. (Chapter 3)

MODELING

Understanding and predicting aviation-induced cloudiness requires global models designed to simulate aerosol evolution and ice nucleation, evaluated with measurements of aerosols.

Recommendation: As part of a national strategy, NASA should support development and assessment of models for all scales of contrail prediction. These models range from wake vortex to global climate to contrail plume to ice supersaturation forecasting. (Chapter 4)

CONTRAIL FORECAST AND VERIFICATION

Reliable prediction of ISSRs with the spatiotemporal requirements sufficient for informing contrail avoidance strategies is likely possible and feasible with evolution of current models and sufficient humidity and temperature data at flight level. Current forecasts of ISSRs and persistent contrails are limited by (1) lack of flight-level in situ humidity data for constraining ISSR locations at flight altitudes, (2) poor parameterizations of cloud physics in weather forecast models that do not permit ice supersaturation, (3) low vertical resolution in the upper troposphere, and (4) lack of adequate models for ice nucleation on existing or aircraft exhaust particles. Current attempts at operational forecast of individual flight or system-level contrail locations and radiative effects are insufficient for making robust net climate impact-reduction mitigation decisions with sufficient skill scores.

New machine learning methods for contrail identification from satellite imagery are showing great promise, but databases are currently fragmented, use different methods, and are not global. Low Earth orbit and geostationary satellite data can be used with aviation location data and machine learning to improve model accuracy and are vital for model validation and testing. Verification of contrail predictions requires good observations of contrails globally. This requires use of multiple satellites and ground-based cameras in an open-source setting.

Recommendation: NASA should support development of a global contrail observing system as a foundation for research, analysis, and future verification. (Chapter 5)

The manner in which Earth science is conducted is evolving. For example, there are private parties exploring the deployment of a specific constellation of satellites to observe contrails. NASA is in a unique position to advise these efforts to ensure that they are best architected and that resulting data are accessible to maximize utility for the broader community. A contrail prediction system requires a forecast model to predict ice supersaturation, which necessitates accounting for the degree of ice supersaturation and improved vertical resolution. New observations for humidity and even contrail imagery would also improve forecasting ISSRs and contrails.

There are several critical priorities to enable contrail mitigation in the near term with higher certainty. First are high-vertical-resolution forecasts, which require improving weather forecast models. Second is estimating high relative humidity regions and ISSRs, which requires better in situ humidity observation, probably from commercial aircraft. It would be best to focus on more certain contrails with known high impact. High-impact contrails have large positive radiative forcing, which comes from long lifetime, high relative humidity, and certain conditions, like nighttime over warm surfaces.

OPERATIONS

The global policy landscape and accuracy of contrail forecast/validation tools are evolving rapidly and are trending toward more widespread adoption of operational avoidance efforts over the coming decades.

Even while the positive warming impact of some contrails is clear, continued research to reduce uncertainties on climate impacts is needed to incentivize airlines and regulators to introduce the additional cost and operational complexity associated with introducing broad avoidance measures.

Current air traffic control systems have difficulty accommodating the rerouting of large traffic volumes for contrail avoidance. Certain operational contrail mitigation concepts, focusing on forecasting the vertical structure of large ISSRs and moving aircraft vertically to avoid high-impact contrails while minimizing extra fuel burn would more readily fit into the current ATC systems.

Trial efforts for rerouting aircraft could be developed with the current air traffic control system to test and evaluate the effectiveness of implementing broad-scale contrail detection and avoidance. To identify specific improvements to the current system, trial tests would

- Illustrate any disruptions or limitations of the current air traffic management system and the current set of models and operations.
- Focus on regions where it is clear that contrails would be high impact (large ISSRs with high amounts of ice at night over warm surfaces).
- Focus on rerouting aircraft vertically (changing flight altitude).
- Engage operators and Air Navigation Service Providers (such as FAA).
- Include significant observations for evaluation by archiving satellite and ground based imagery, as well as use in situ aircraft (including NASA observation aircraft) to measure contrails.
- Test different weather forecast models.

Recommendation: NASA, in collaboration with airline operators and Air Navigation Service Providers, should continue research, development, and operational evaluation of advanced high-altitude air traffic control concepts of operations to enable flexibility to accommodate fuel efficient and contrail avoidance flight trajectories. (Chapter 6)

VISION FOR CONTRAILS RESEARCH

A key objective for the United States is maintaining leadership in the aviation sector, and contrails research is important for the United States' ability to address this emerging aviation issue. The committee has recommended both short- and long-term contrails research activities. The committee's vision for national contrails research is that it will provide sufficient knowledge that can be used by a broad range of government, industry, academic, and international actors. It could ultimately inform a contrails mitigation strategy. The United States is strongly positioned to conduct this research to enable the country to have a competitive advantage both economically and diplomatically, while leveraging the existing body of international, collaborative research in this area. A national research strategy with NASA coordinating with other relevant government agencies and commercial and international partners is vital for demonstrating that the United States will not be left behind in this area.

1

Overview

The cumulative emissions of the air transportation sector, while generating significant societal benefits, currently contribute approximately 2.5 percent of the overall global carbon dioxide (CO_2) emissions (and about 1 percent of total radiative forcing) from human (anthropogenic) activity. Even with advancements in airplane and engine technologies that have reduced the CO_2 intensity of aviation, overall emissions continue to rise as the growth in air traffic outpaces efficiency gains. Over the past decades with significant investment in technologies, annual aviation fuel burn efficiency improvements have averaged 1–2 percent; however, fuel burn and emissions have grown by 4–5 percent annually, leading to a net increase in total CO_2 emissions (Lee et al. 2021 and references therein). The increases in the aviation sector stand out compared to decreasing emissions from other sectors due to phasing out of fossil fuels. The radiative forcing effects on climate of aviation-related CO_2 emissions is well understood with a high degree of confidence. A critical characteristic of anthropogenic CO_2 emissions is their long atmospheric lifetime, persisting for about a century before being reabsorbed through oceanic and atmospheric processes.

Aviation condensation trails (contrails) and their associated aviation-induced cloudiness are, in addition to CO_2, another significant source of aviation climate impact and are assessed to contribute another ~1.5 percent of total anthropogenic radiative forcing of climate (IPCC 1999; Lee et al. 2021). Contrails (Figure 1-1) are created when warm aircraft engine exhaust, laden with water vapor and particulates, interacts with a cold ambient atmosphere with humidity higher than the saturation vapor pressure over ice ("supersaturated" with respect to ice). Under conditions defined by the Schmidt–Appleman criterion (Appleman 1953; Schumann 1996), which describes mixing of warm and humid jet exhaust with surrounding ambient air, the water vapor in engine exhaust will condense on the particulates in the exhaust plume and freeze to form a contrail (Figure 1-1, lower left). If the temperature is cold enough and the atmosphere is supersaturated with respect to ice (which can be frequent in clear skies in the upper troposphere), then a contrail can form and persist, beyond 10 minutes. These persistent contrails are of concern for climate impact.

Contrails influence climate by altering Earth's energy balance, contributing both warming and cooling effects. Their overall effects are assessed through the concept of radiative forcing, which measures the change in energy flux at the top of the atmosphere caused by contrails. This includes both the heat trapped (longwave effect) and the sunlight reflected back into space (shortwave effect).

Contrails reflect daytime incoming solar radiation and also absorb and re-radiate outgoing infrared radiation to Earth and space (Figure 1-1, upper left). Since clouds are white, and generally brighter than the surface below, they will generally cool the surface in the solar or "shortwave" bands. Their maximum cooling effect is

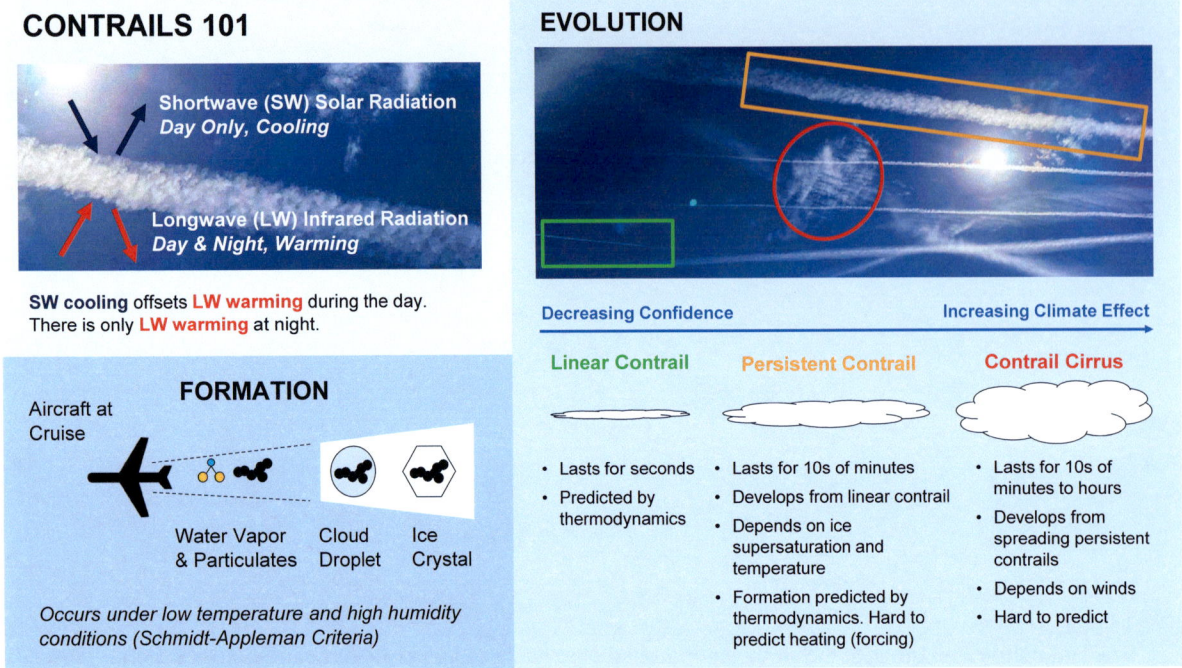

FIGURE 1-1 Schematic of key contrail concepts. *Upper left:* Contrail radiative effects. Contrails cool by reflecting solar radiation back to space (white arrow); contrails warm by absorbing infrared radiation and re-emitting it back to Earth (red arrow). *Lower left:* Contrail formation occurs when water vapor condenses on particulates (often soot) to form a cloud droplet and then freezes at low temperature and high enough humidity. *Right:* Contrail evolution from short-lived linear contrails (green) to persistent contrails (yellow) to contrail cirrus (red).
SOURCE: Airplane icon courtesy of iconsDB.com.

over a dark surface (ocean), with almost no cooling effect over snow, ice, other clouds, or at night. In the infrared ("longwave" bands), contrails are colder than the underlying surface (or clouds) below, and they trap energy and re-radiate or scatter it at a colder temperature to space (and back down to Earth), warming the surface. The effect is largest over warmer surfaces and occurs day or night.

The net warming versus cooling effect of contrails depends on many factors, including ambient temperature, ice mass, crystal size and shape (determining the optical thickness), as well as the location of the Sun and the underlying Earth surface or cloud brightness and temperature. Another key attribute of contrails is that, if they form, they can have a persistence on the order of minutes to hours (Figure 1-1, right). The integrated effect of an individual contrail depends on its lifetime and the factors above. On balance, the net effect of all contrails (as in general with natural cirrus ice clouds) is to warm the planet, although some individual contrails can locally cool the planet.

The key for predicting where a persistent contrail will form that is significant for climate is the ability to predict where the atmosphere is supersaturated with respect to ice, known as ice-supersaturated regions (ISSRs). ISSRs occur frequently in the upper troposphere at commercial transport airplane cruising altitudes (Gettelman et al. 2006). If a flight occurs through an ISSR and the Schmidt–Appleman criterion is met (cold enough temperature and high enough humidity), then a contrail will form and may persist. Persistent contrails in an ISSR take up water from the ambient atmosphere, so that most of the ice mass in a contrail comes not from the aircraft exhaust, but from ambient water vapor. Persistent contrails can last from tens of minutes to hours, and longer-lived contrails can further evolve from linear features into "contrail cirrus" (Figure 1-1, right). To determine the aggregate radiative effect of many contrails from operation of the entire aviation fleet, individual contrails are aggregated across global flights and applied over the area of the planet. Individual contrails last only a short time; however, the global aircraft fleet flies continuously and creates contrails continuously, thus overall creating a warming effect, which is usually assessed on an annual basis, with variations due to the seasonal evolution of the ambient atmosphere and seasonal changes to aircraft routing.

The climate impact of contrails, other aerosols, and greenhouse gases was first estimated through *instantaneous radiative forcing* (RF_{inst} or RF), which is calculated as the net change in radiative balance (solar heating minus thermal cooling) in watts per meter squared (W/m^2) at the top of the atmosphere caused by an instantaneous change in atmospheric composition (e.g., occurrence of a contrail, or increase in mean CO_2 abundance). The assumption in the early generations of Intergovernmental Panel on Climate Change (IPCC) climate assessments was that the global mean warming was linearly proportional to the persistent RF change, and thus RF became the metric to compare climate change across greenhouse gases and aerosols. As climate science evolved, it became clear that for some types of perturbants, the RF did not accurately predict warming because the atmosphere develops short-term responses to the RF_{inst}. Current IPCC climate assessments use *effective radiative forcing* (ERF), which accounts for the RF change after the atmosphere has adjusted to the perturbation (e.g., via altered temperatures and/or clouds). The difference between RF and ERF varies by forcing agent (see Figure 1-3 later in this section and discussion in Lee et al. [2021]).

The radiative forcing effects of contrails versus other aviation emissions spans orders of magnitudes relative to their persistence. Contrails create a very-short-lived radiative forcing (hours or days); nitrogen oxides (NOx) emissions perturb the greenhouse gases CH_4 and O_3 for months to decades; while CO_2 emissions decay on timescales of decades to centuries or longer. The impact of these differences in persistence on climate change metrics is discussed in Box 1-1.

ISSRs are the primary condition for the creation and persistence of contrails. Due to the uncertainties in forecasting ISSRs, there is significant uncertainty in developing effective mitigation strategies based on contrail avoidance. The sources of uncertainty in estimating contrail formation and corresponding contrail radiative effects are illustrated in Figure 1-2.

The observation and prediction of ISSRs is challenging, but given a flight through an ISSR, it is typically well determined that a persistent contrail will form. The resulting initial ice crystal size distribution formed after a few aircraft lengths of wake affects the ultimate contrail optical properties (Figure 1-2) and depends on the total particulates emitted from the engine (both engine produced and the ambient particulates that pass through the engine or mix with the contrail plume). The concentration of particles varies significantly by aircraft engine type, engine power setting, and jet fuel chemical composition. Particulates are generally soot and other non-volatile particulate matter, as well as some condensed volatiles (such as sulfates or organics). Because of the high degree of water supersaturation that occurs initially in the plume, almost all particulates above a certain size nucleate water droplets that subsequently freeze. Observations exist for both aircraft particulate emissions and of initial contrail ice particle size distributions, which are used for calibrating model simulations of contrails.

The ERF of a contrail depends on the distribution of ice crystal size, mass density, and particle shape. These properties determine the optical depth of the contrail over time (Figure 1-2). The initial ice crystal size distribution evolves as the contrail takes up ambient water. The speed and amount of water vapor uptake depends on the degree of supersaturation: an ISSR with a relative humidity (RH) of 120 percent over ice will result in more water vapor uptake (and more ice mass) than an ISSR with 105 percent RH over ice. Over time, the contrail evolves in the atmosphere, undergoing shearing due to atmospheric processes and with its ice crystals sedimenting out as they grow. If the ambient atmosphere then becomes subsaturated with respect to ice, the contrail will begin to sublimate, diminish, and disappear.

Observations of contrails exist from various observational platforms and instrument types: ground based, airborne in situ, airborne remote sensing, and space-based remote sensing. The evolution of a contrail (horizontal and vertical extent) is determined by atmospheric conditions, including winds, and wind gradients that can shear out a contrail. Observations and data sets from satellites and the ground exist for contrails for age, extent (size), and lifetime. The total radiative effect of a contrail is determined by the following: its optical depth over space and time, its location relative to the Sun, surface albedo, or reflectance (to determine net shortwave reflection), and the underlying surface temperature (to determine the net longwave absorption). The radiative effect of a contrail also

> **BOX 1-1**
> **A Temporal View of Climate Impact Metrics**
>
> From the scale of shifting an individual flight to rerouting the entire aviation fleet, reducing contrail impacts requires a trade-off between eliminating contrails and burning slightly more fuel, thereby emitting more CO_2 and NOx on the modified route. In terms of climate impact trade-offs, there are effective radiative forcing (ERF) estimates of the different emissions/impacts (contrails, CO_2 and NOx) from models. As ERF is expressed as a rate (radiative effect per unit area and unit time, W/m^2), it is necessary to understand the timescales over which effects occur to compare impacts. The persistence of ERFs from aviation span a large range of timescales, as illustrated in Figure 1-1-1, where 1 year of aviation emissions is illustrated schematically. The ERF persistence varies as follows: a century or more for CO_2; 1 month and 10 years for NOx (two different effects); and hours for contrails. Consequently, the total effect of contrails or NOx emissions from an activity (a flight, or a year of flights, also called an "impulse") tends to be immediate, while aviation CO_2 and the 10-year NOx effects on methane act over a much longer period of time than just the year of activity. These ERF curves must be integrated over time to calculate the total climate impact. Note that Figure 1-3 from Lee et al. (2021) represents cumulative emissions to date, so is different than Figure 1-1-1.
>
>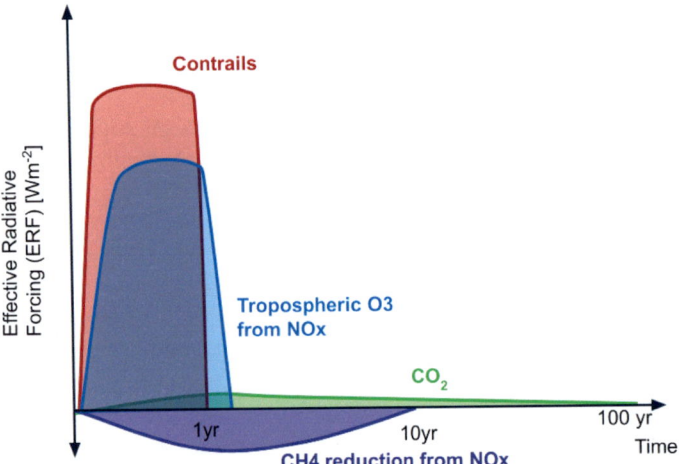
>
> **FIGURE 1-1-1** Schematic diagram of aviation effective radiative forcing (ERF) from 1 year of fleet aviation operations. Contrail ERF (red) turns on within hours of operation and shuts off similarly when flights stop. CO_2 ERF (green) has a smaller initial impact, accumulates almost linearly over the year, and then lasts for many years, decaying on a century-long scale. NOx ERF has two components: (1) increases in tropospheric O_3 (light blue) that build up over the first month and then decay rapidly when emissions cease and (2) reductions in CH_4 (dark blue, negative ERF, cooling) that build up over the year of operations and then decay on a decade-long scale. Note that the time axis is logarithmic, and the ERF axis is only approximately to scale.

depends on other clouds (including contrails) below or above it. Two similar contrails with a different position relative to the Sun and with a different albedo below will have different radiative effects.

Several studies have addressed quantifying the overall radiative effect of contrails and understanding their uncertainties. The studies took various approaches and utilized different methodologies. However, they were all constrained by observations and theory and considered the chain of processes in Figure 1-2, to estimate the contrail optical depth and lifetime to determine the radiative forcing of resulting contrail cirrus, and to

Many climate change metrics have been defined for comparing disparate sources of climate forcing, and each has a different purpose depending on the policy goals (e.g., Forster et al. [2021]; Myhre et al. [2013]). The Global Warming Potential (GWP) was the original metric for comparing emissions of different greenhouse gases (GHGs): carbon dioxide, methane, nitrous oxide, and chlorofluorocarbons. GWP has been adopted as the standard by the governments under the United Nations Framework Convention on Climate Change. For long-lived, well-mixed GHGs, their ERF does not depend on where they were emitted in the lower atmosphere. GWP for a GHG is the ratio of the warming to that caused by the same mass emission (kg) of CO_2.

The timescale selected for climate metrics is a decision based on policy choices, not just the physics of the atmosphere. When defined in the late 20th century, climate change research was concerned with 100-year timelines, so GWP-100 (100-year time integration) was used. The 2015 Paris Accord shifted the focus to the 2050 time frame, and thus there is more focus now on 20-year timescales.

In addition, new metrics, such as the Global Temperature Potential (GTP), are being used. GTP integrates the ERF effects of emissions over time to calculate the change in temperature at the end of the GTP time frame. Like GWP, GTP is a ratio of the effect relative to the CO_2 impact on a kilogram-per-kilogram emission basis. GWP and related metrics like GTP are based on ratios of 1 kg emissions and thus cannot directly apply to contrails, or indirect GHGs like NOx that act by altering other species, particularly ozone and methane.

For contrails, the climate effect is the annual mean over time of the ERF from an "activity" (either an individual flight, or 1 year of operations of the civil fleet).[a] Because the contrail effects do not last much beyond the time of the flight (or the year of operations), the integrated ERF remains constant over 20- or 100-year time horizons. For CO_2 emitted from aircraft (one or a year of operations), the instantaneous ERF largely continues out to 100 years or beyond (Figure 1-1-1), and so the integrated ERF increases as the time window is longer. Thus, contrail mitigation is relatively more important with respect to CO_2 emissions when using a shorter time horizon. The comparison is not readily available from the ERF bar chart of Lee et al. (2021) because the CO_2 ERF includes past accumulation but not future effects.

A common desire among users of climate change metrics is to declare a Social Cost of Carbon (SCC) that measures in dollars long-term damage caused by emissions of 1 ton of CO_2. Federal agencies often use the SCC estimates to place a value on the climate impacts of rulemakings. The economic costs (e.g., added energy use) can be evaluated and there are many SCC values, especially in the economics literature. Because "cost-benefit analysis remains limited in its ability to represent all avoided damages from climate change" (IPCC 2023), estimates of the SCC are highly uncertain and almost certainly biased low.

---

[a] The "civil fleet" includes commercial, non-military government aircraft, and business aircraft, and the latter two categories are a small percentage of the civil fleet. Most aviation climate studies discuss the civil fleet.

estimate the ERF of the overall aviation fleet. Conceptually, when an ISSR is present at given temperatures, contrails will form and persist, but forecasting and simulating these regions is challenging. Particulate emissions can vary widely, as determined by measurements from different aircraft engine types and fuels, as well as the resulting near-field ice crystal sizes (ice number reflects engine particle emissions). Observations of contrail extent and lifetime from the ground and satellites provide distributions for comparison. All these observations are used to constrain model estimates: some are bottom-up models that track each individual

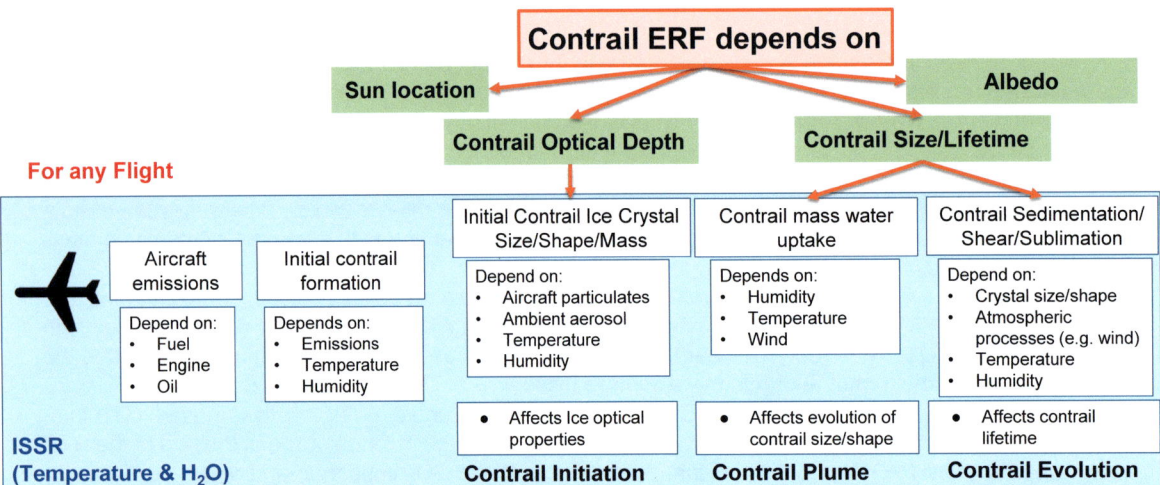

FIGURE 1-2 Understanding contrail radiative forcing and what drives uncertainty, from fuel to formation to plume and contrail evolution.
NOTE: ERF, effective radiative forcing; ISSR, ice-supersaturated region.
SOURCE: Airplane icon courtesy of iconsDB.com.

contrail using a static historical atmosphere, and some are climate models that initiate contrails that evolve in a simulated atmosphere.

Each methodology has its own uncertainties and sensitivities, which are assessed in various publications. Recent assessments (Lee et al. 2021) have tried to harmonize these studies to have the same assumptions and provide a best global contrail ERF estimate of about 0.06 W/m^2, with large uncertainties of about ±70 percent (Figure 1-3). The uncertainty range for the total fleet is limited because there are observational constraints on contrails (lifetime, geometric extent, crystal size, optical depth) and some contrasting effects (big particles have more mass but sediment faster). The ERF from cumulative aviation CO_2 emissions is 0.03 W/m^2 (and growing slightly) with uncertainties of about 17–19 percent.

Aviation emissions of nitrogen oxides (NOx) are another important non-CO_2 source of climate forcing (Figure 1-3), and these impacts are also relatively short lived compared to CO_2. NOx is generated internally to the engine where the high temperatures of combustion cause reactions of atmospheric nitrogen and oxygen. Driven by atmospheric chemistry, these NOx emissions lead to short-term increases in tropospheric ozone (O_3), which is itself a greenhouse gas. The NOx-related perturbation of O_3 has a nominal ERF impact on the same order as that of CO_2. However, other atmospheric chemistry processes involving NOx lead to reductions in atmospheric methane, another major greenhouse gas, which offsets some of the impact of ozone. Overall, these opposing effects lead to a current net ERF impact of NOx with a best estimate that is 50 percent lower than that of CO_2 and 70 percent lower than that of contrails (Figure 1-3). However, these offsetting effects also result in a relatively large uncertainty range for net NOx ERF; that is, from −97 to +66 percent of the central estimate. Therefore, the net ERF effects of NOx are correspondingly assessed with a low confidence level.

Quantifying the climate impact of aviation emissions, particularly for contrails and NOx emissions, has been the focus of many researchers, with several recent assessments of this work (Lee et al. 2021, Figure 1-3, 2023). The assessment of aviation CO_2 ERF in Figure 1-3 was performed over the period of 1940 (when aviation contrails were first observed) through 2018 by normalizing the methodologies and results across many different studies such that they could be used to develop consensus results. Short-term (NOx and contrail) effects represent a steady state from the 2018 fleet. The results show that the largest non-CO_2 warming effect of aviation is due to contrails and their associated aviation-induced cloudiness (which is discussed further in Chapter 2). The climate

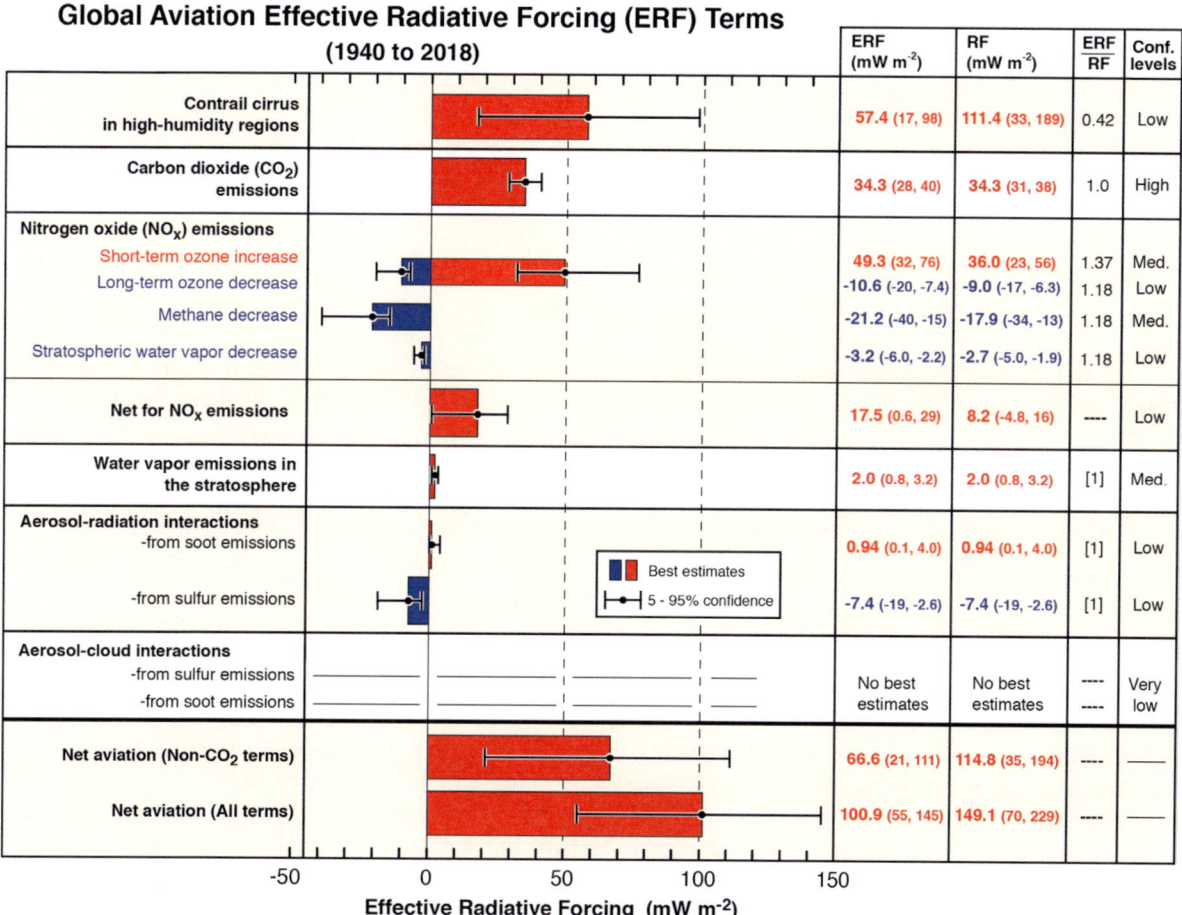

FIGURE 1-3 Effective radiative effects of aviation, by component/emissions species, measured in milliwatts per square meter (mW m^{-2}). 1000 mW m^{-2} = 1 W m^{-2}. "Contrail-cirrus in high humidity regions" refers to persistent contrails.
SOURCE: Reprinted from D.S. Lee, D.W. Fahey, A. Skowron, et al., 2021, "The Contribution of Global Aviation to Anthropogenic Climate Forcing for 2000 to 2018," *Atmospheric Environment* 244:117834, with permission from Elsevier, © 2020 Elsevier.

warming impact of contrails, measured as ERF, has been estimated to be 67 percent greater than that of CO_2 in 2018. However, the ERF impact of contrails has significant uncertainty, that is, ±70 percent, due to uncertainties in underlying processes and methods, and therefore their impact has been assessed with a low confidence level. Validation of models and the diversity of approaches provides some robustness to the sign of the contrail impact: the effects are on aggregate a warming effect. Combining the CO_2 and non-CO_2 effects from contrails and NOx yields an overall ERF from aviation of about 0.1 W/m^2, or about 4 percent of total anthropogenic forcing.

The assessment of the ERF warming impact of contrails by Lee et al. (2021) has been estimated based on several studies, employing a range of methodologies, and with efforts to harmonize the results. Typically, these studies use a combination of contrail observations as well as weather data and/or model simulations. Model studies of contrail climate effects are typically calibrated by observations (e.g., observed contrail ice crystal sizes, contrail sizes) and use uncertainty (sensitivity) analysis to quantify the possible range of estimates. The individual studies were then harmonized based on their assumptions, and the uncertainty was determined from the statistical combination of the reported uncertainty from each study.

Finding: The current overall aviation impact on climate, comprises ~4 percent of all anthropogenic climate forcing.

Finding: The effect of aviation due to contrails and aviation-induced cloudiness is comparable to aviation CO_2 with the timescale of impact determining which is larger. Aviation NOx emissions cause two separate impacts, each of which is comparable to aviation CO_2 in the absolute sense, but they have opposite signs and thus the net positive forcing is about half that of aviation CO_2.

Finding: In aggregate, there is confidence that contrails create a net warming effect. The uncertainty in this climate forcing is large but does not include net cooling.

Strongly warming contrails occur in a relatively small percentage of all flights (Teoh et al. 2024). It is estimated that most (~80 percent) flights have negligible or no impact on climate forcing. Only a small percentage of flights and flight distances result in the majority of overall warming impact of contrails, and thus contrail avoidance must include the skill in predicting the few flights that will have the largest contrail warming effect.

Finding: A relatively small percentage of flights account for the majority of the contrail warming effect.

REPORT OUTLINE

This report will discuss the chain of contrail effects from aircraft to contrail evolution and provide recommendations on where NASA can focus on research to accelerate understanding of the contrail formation, life cycle, and options for mitigation of contrail climate effects. Figure 1-4 illustrates this schematically.

Contrail evolution and radiative forcing (as well as aviation-induced cloudiness) are affected by the emissions of aircraft engine particulate matter (Chapter 2). The engine combustion characteristics and the fuel composition strongly influence the particulate concentrations and size distributions. More observations are necessary to assess the impact of the chemical composition of alternative jet fuels on total engine particulate emissions. Understanding contrail effects first requires good measurements of the atmosphere (Chapter 3). Ambient temperature and humidity control the formation and persistence of contrails, and higher-quality observations are necessary to

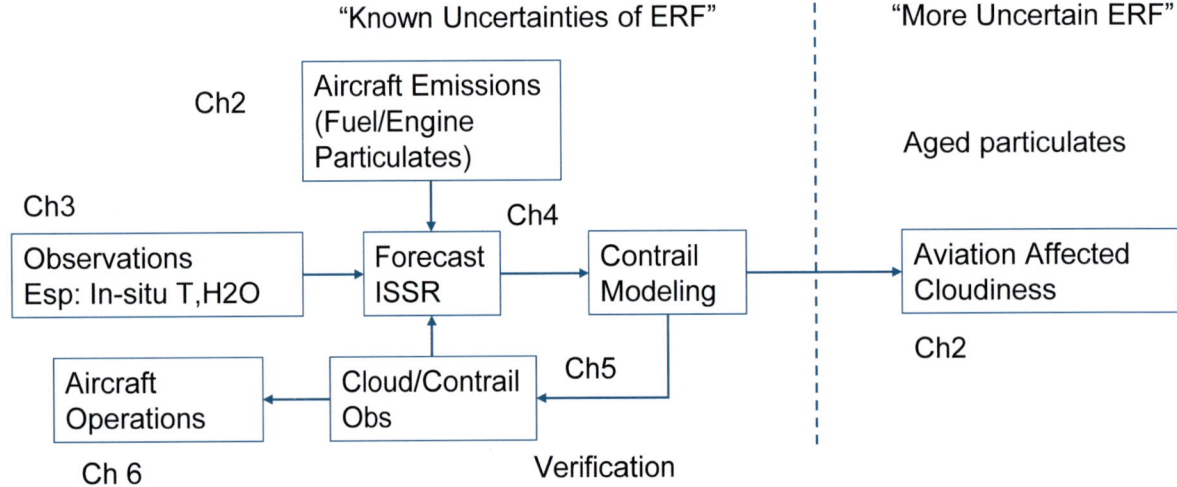

FIGURE 1-4 Roadmap to understand and potentially reduce contrail climate effects. Chapter 2 will discuss emissions from aviation. Chapter 3 discusses observations, Chapter 4 discusses modeling contrails and ice-supersaturated regions (ISSRs), Chapter 5 discusses forecast systems and verification, and Chapter 6 aircraft operations.

improve ISSR forecasting. These measurements need to be accurate and with high vertical resolution regarding altitudes and flight levels, which could be achieved via increased deployment of sensors on board commercial aircraft. The most important part of predicting warming contrails for a specific set of aircraft is predicting ISSRs with models (Chapter 4) and then also simulating contrail initiation and evolution. Verification of forecasts and observing contrails in real time is also critical for assessing any changes to aircraft routings, and this can be done with constrained and data-driven models, as well as satellite observations of contrails in forecast systems (Chapter 5). Finally, to try to mitigate the effects of contrails, aircraft operations (discussed in Chapter 6) will have to incorporate validated contrail predictions based on observations (discussed in Chapter 3), aircraft engine type and fuel (discussed in Chapter 2), and ISSR and contrail evolution forecasts (Chapter 4) that are suitably verified (Chapter 5). This comprehensive research plan has a few places where NASA has unique capabilities to advance science and operations. Additional findings and recommendations are in the following chapters (Figure 1-5).

Mitigating contrail effects on climate requires the engagement of many stakeholders, including scientists, regulators, airframers, engine manufacturers, and operators. NASA is in a unique position as a non-regulator that touches a number of parts of the aviation-climate space. The agency has both resources and a reputation that positions it to take a leading role in this research to attain U.S. international leadership and competitiveness.

FIGURE 1-5 Contrail over Manhattan, late 2024.

REFERENCES

Appleman, H.S. 1953. "The Formation of Exhaust Condensation Trails by Jet Aircraft." *Bulletin of the American Meteorological Society* 34(1):14–20. https://doi.org/10.1175/1520-0477-34.1.14.

Forster, P., T. Storelvmo, K. Armour, W. Collins, J.-L. Dufresne, D. Frame, D.J. Lunt, et al. 2021. "The Earth's Energy Budget, Climate Feedbacks, and Climate Sensitivity." Pp. 923–1054 in *Climate Change 2021: The Physical Science Basis. Contribution of Working Group I to the Sixth Assessment Report of the Intergovernmental Panel on Climate Change*, V. Masson-Delmotte, P. Zhai, A. Pirani, S.L. Connors, C. Péan, S. Berger, N. Caud, et al., eds. Cambridge University Press. https://doi.org/10.1017/9781009157896.009.

Gettelman, A., E.J. Fetzer, F.W. Irion, and A. Eldering. 2006. "The Global Distribution of Supersaturation in the Upper Troposphere from the Atmospheric Infrared Sounder." *Journal of Climate* 19:6089–6103.

IPCC (Intergovernmental Panel on Climate Change). 2023. "Climate Change 2023: Synthesis Report Summary for Policymakers." Contribution of Working Groups I, II and III to the Sixth Assessment Report. https://doi.org/10.59327/IPCC/AR6-9789291691647.

Lee, D.S., D.W. Fahey, A. Skowron, M.R. Allen, U. Burkhardt, Q. Chen, S.J. Doherty, et al. 2021. "The Contribution of Global Aviation to Anthropogenic Climate Forcing for 2000 to 2018." *Atmospheric Environment* 244:117834.

Lee, D.S., M.R. Allen, N. Cumpsty, B. Owen, K.P. Shine, and A. Skowron. 2023. "Uncertainties in Mitigating Aviation Non-CO_2 Emissions for Climate and Air Quality Using Hydrocarbon Fuels." *Environmental Science: Atmospheres* 12(3):1693–1740.

Myhre, G., D. Shindell, F.-M. Bréon, W. Collins, J. Fuglestvedt, J. Huang, D. Koch, et al. 2013. "Anthropogenic and Natural Radiative Forcing." *Climate Change 2013: The Physical Science Basis*. Contribution of Working Group I to the Fifth Assessment Report of the Intergovernmental Panel on Climate Change.

Schumann, U. 1996. "On Conditions for Contrail Formation from Aircraft Exhausts." *Meteorologische Zeitschrift* 5(1):4–23. https://doi.org/10.1127/metz/5/1996/4.

Teoh, R., Z. Engberg, U. Schumann, C. Voigt, M. Shapiro, S. Rohs, and M.E.J. Stettler. 2024. "Global Aviation Contrail Climate Effects from 2019 to 2021." *Atmospheric Chemistry and Physics* 24(10):6071–6093.

2

Aircraft Engine Emissions

Chapter 1 noted that aircraft particulate emissions initiate formation of aviation-induced cloudiness. The purpose of this chapter is to explain the factors that influence these aircraft engine emissions, as well as the broader safety and operational factors that aircraft engines must be designed to address that interact with factors that influence emissions. It is organized by first introducing high-level issues, then addresses influences of combustor technology, ambient and engine operating conditions, and sustainable aviation fuels (SAFs). It then discusses effects of lubrication oil venting and how the temporal evolution of engine emissions influence contrail formation.

Combustion of hydrocarbon fuels leads to emissions of particulates and other precursors that serve as condensation nuclei for contrails (Figure 2-1). Combustor technology has a significant influence on these emissions and, hence, on contrail-forming tendencies. It is possible to significantly reduce these emissions with advanced combustion technologies and, in turn, decrease but not eliminate contrail-forming tendencies. The reason they do not eliminate contrail formation is that, even if combustion particulate emissions could be eliminated (such as with hydrogen as a fuel), ambient atmospheric particles, as well as trace amounts of oil vapor and aerosols vented from the engine lubrication systems, also may serve as condensation nucleation for contrail formation. In addition, the chemical composition of the fuel has a significant influence on contrail-forming tendencies. Conventional jet fuels contain sulfur and aromatic hydrocarbons, both of which are important sources of particulates relevant for contrail formation. Some SAFs contain fewer or no sulfur and/or aromatic compounds than conventional jet fuels and accordingly yield significantly decreased particulate emissions and contrail-forming tendencies, while other SAFs have sulfur and/or aromatic compounds at levels consistent with the current Jet A specification (so called drop-in fuels). Each of these items is further detailed in the rest of this chapter.

GENERAL COMBUSTOR EMISSIONS

Aircraft Engines

Aircraft engine combustors enable the conversion of chemical energy stored in jet fuels to thermal energy via exothermic chemical reactions. Jet fuels are a combination of hydrocarbons that meet the technical and certification requirements for aviation. Key performance metrics of the combustor are to fully burn the fuel ("combustion efficiency"), operate stably over the entire operational range of the engine power ("operability"), have a reason-

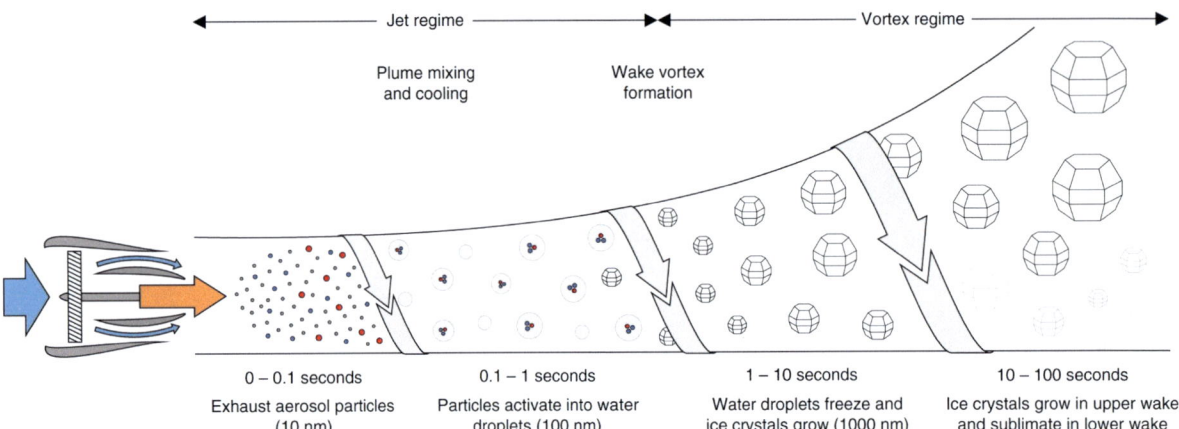

FIGURE 2-1 Contrail formation and evolution.
SOURCE: B. Kärcher, 2018, "Formation and Radiative Forcing of Contrail Cirrus," *Nature Communications* 9:1824, https://doi.org/10.1038/s41467-018-04068-0. CC BY-NC-ND 4.0.

ably uniform exit temperature profile (i.e., mild temperature gradients in the radial and azimuthal directions), and have low pollutant emissions.

The combustion process yields the main product species of gaseous CO_2 and H_2O, which are greenhouse gases. Long-lived CO_2 is of primary importance for climate, while the impact of water vapor depends on whether it is emitted in the troposphere, where it precipitates relatively quickly from the atmosphere, or the stratosphere, where its longer lifetime can be climate relevant. This process also produces primary pollutants, such as soot or non-volatile particulate matter (nvPM) and sulfur dioxide (SO_2). Soot particles are generated during the combustion process due to imperfect local mixing of available jet fuel droplets and available air. If the fuel contains sulfur, SO_2 is produced during the combustion process and then is further oxidized to form sulfuric acid, which, along with unburned semi-volatile organic compounds, contributes to the formation of volatile particulate matter (vPM) downstream of the engine nozzle exit. The combustion process also produces nitrogen oxides (NOx), which arise due to chemical reactions of nitrogen and oxygen in the air across the flame. Combustor designs that decrease the residence time of fluid particles after the flame, or that minimize local hot spots due to imperfect mixing, lead to decreased NOx emissions. Additional potential combustor emissions that influence combustor design include carbon monoxide (CO) and unburned hydrocarbons (UHCs).

Aircraft emissions, specifically nvPM emissions and NOx, are regulated via the Environmental Protection Agency Integrated Pollution Parameter metric standards. These standards are set by the Committee on Aviation Environmental Protection of the United Nations' International Civil Aviation Organization (ICAO) for the landing/take-off cycle conditions, versus in-flight or cruise conditions. As a result, aircraft combustor systems are optimized to meet these emissions regulations, while balancing other critical design criteria, such as efficiency, operability, heat release, exit temperature uniformity, and durability. The combustor design optimization process considers the air flow around and through the combustion system, the turbulent mixing effectiveness of fuel and air to reduce soot emissions, the mixing of air downstream of the flame to meet exit temperature quality requirements, and the use of cooling air to maintain combustor liner wall temperatures to ensure component life requirements. Modern engines have dramatically reduced the soot particle emissions over the past half century, as evidenced by ever lower smoke numbers being reported in the ICAO Aircraft Engine Emissions Databank.[1] Mass and number emissions indices for nvPM have also been included in the Databank since 2020, which will allow for a more quantitative assessment of engine particle emissions improvements in the future; however,

---

[1] See European Union Aviation Safety Agency, "ICAO Aircraft Engine Emissions Databank," https://www.easa.europa.eu/en/domains/environment/icao-aircraft-engine-emissions-databank, accessed December 1, 2024.

these emissions indices are reported at sea-level static thrust conditions that may not be representative of cruise conditions. Recent research has attempted to extrapolate on-wing non-volatile particle emissions measurements made on the ground to cruise altitude measurements with some success for older technology engines burning conventional petroleum-based fuels (e.g., Ahrens et al. 2023; Durdina et al. 2017; Peck et al. 2013; Petzold et al. 1999; Stettler et al. 2013). It remains to be shown whether these ground-to-cruise correlation methods are applicable to more modern engine combustor technologies as well as sustainable or non-drop-in fuels. Similarly, it is also known that engine particle emissions are impacted by the maintenance history of a specific engine. Certification data in the ICAO Databank are for new engines and may not reflect the emissions of in-service engines across the fleet.

Aircraft nvPM and vPM, as well as ambient aerosol particles, provide the nuclei that enable water condensation and ice crystal growth under water-supersaturated conditions at cruise altitudes and, as such, are critical parts of the overall persistent contrail issue.

Finding: Contrail ice crystals form on volatile and non-volatile particles emitted from the aircraft engines as well as ambient particles in the upper troposphere.

The number of ice crystals and their optical properties influence the optical depth, persistence, radiative forcing, and subsequent climate impacts of the contrail. These contrail microphysical properties are determined, in part, by the aircraft engine particulate size distribution, which often comprises two size modes: a smaller nucleation size mode dominated by numerous volatile particles (number mode diameter ~1–20 nm) and a slightly larger soot mode dominated by nvPM (number mode diameter ~20–40 nm). For aircraft engines with rich-burn, quick-quench, lean-burn (RQL) combustors, the number of emitted soot particles ($>\sim 10^{14}$ kg-fuel^{-1}) thermodynamically determines the number of contrail ice crystals formed. Meanwhile, the more numerous, but smaller, nucleation-mode particles do not significantly impact the contrail microphysical properties due to the Kelvin effect. This "soot-rich regime" is shown in Figure 2-2 as a proportional scaling between ice crystal number and soot particle number. For these combustion systems, novel component design choices that yield lower nvPM emissions are expected to yield fewer contrail ice particles, lower optical depth, and reduced radiative forcing to first order.

Finding: Aircraft engine particulate emissions influence contrail properties and resultant radiative forcing.

Some modern aircraft engines include combustion systems that yield jet exhaust conditions in the "soot-poor regime," with soot emissions up to three orders of magnitude lower than combustion systems that operate in the soot-rich regime. The lean-burn combustion technology in some current engines yields emissions in the soot-poor regime, while some RQL combustion technologies in other aircraft engines yield jet exhaust conditions in the transition region between soot-rich and soot-poor regimes. In the soot-poor regime, the number of emitted soot particles ($<\sim 10^{13}$–10^{14} kg-fuel^{-1}) is insufficient to quickly uptake the condensing water vapor from the engine exhaust, and the plume water supersaturation may cause ambient aerosols and the smaller, nucleation-mode particle emissions to grow to become contrail particles. For atmospheric temperatures well below the contrail formation threshold, this means that further reductions in soot particle emissions (assuming constant emissions of nucleation mode or ultrafine particles from sulfur compounds or engine oil) might increase the number of contrail ice crystals formed (left side of Figure 2-2). However, the current number of observations within the soot-poor regime is very limited and other engine design and fuel considerations beyond the combustor (e.g., lubrication oil venting, fuel sulfur content) will be important in this regime.

The main sources of volatile particulate matter are sulfur oxides (SO_x), condensable organics from incomplete combustion, and vented oil vapor and aerosols. The location of the oil vapor exhaust vent (e.g., near the core flow, bypass flow, or outside the nacelle) as well as the concentration of soot particles impact how the emitted semi-volatile oil vapor and aerosol contributes to vPM. If the oil emissions are in a cold flow (bypass duct or outside the nacelle), a distinct larger oil particle size distribution mode will be present and contribute (relatively) few condensation nuclei for contrail ice crystal formation. If the oil vents into the hot core flow, the oil will be vaporized and recondense onto existing soot particles or nucleate new particles.

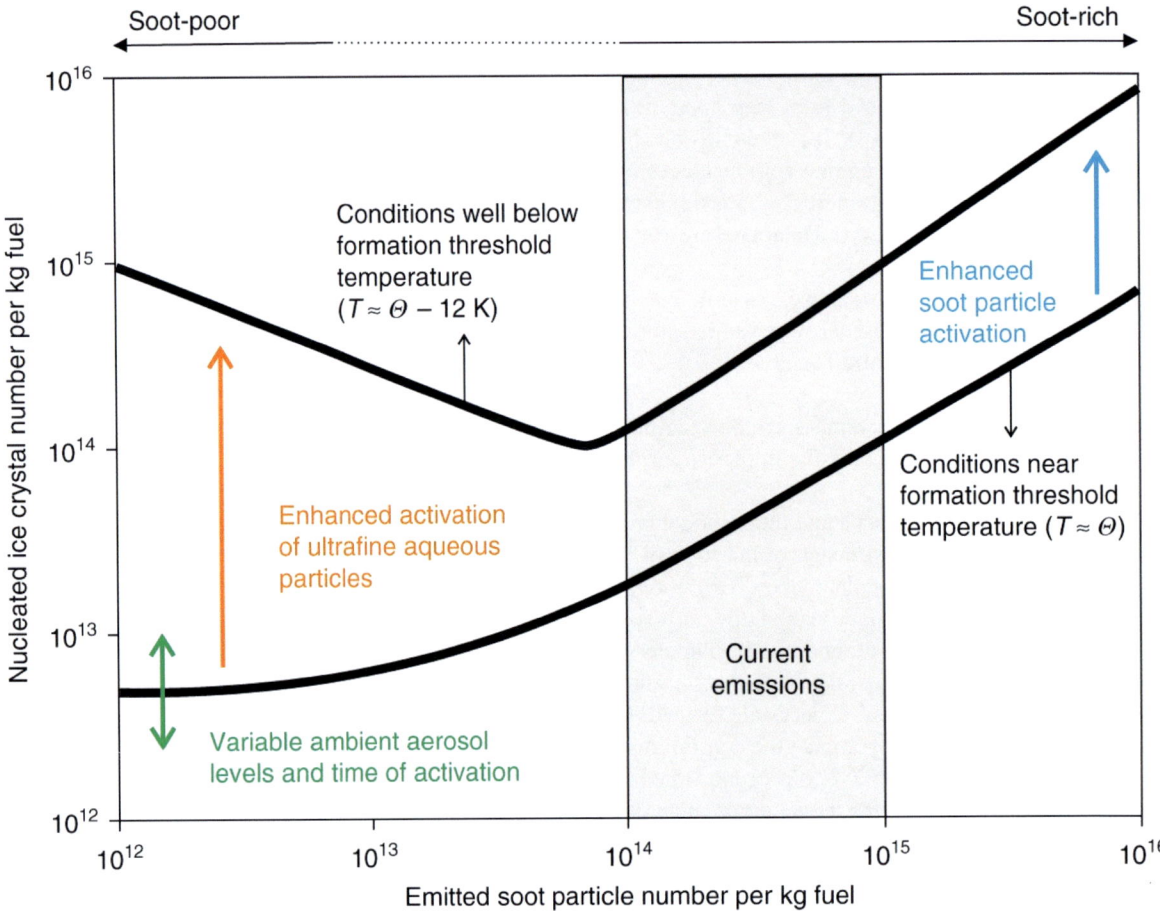

FIGURE 2-2 Nucleated ice crystal number by jet engine exhaust regime. Soot-rich regime refers to emitted particle numbers per mass of fuel greater than 10^{14}. Soot-poor regime refers to emitted particle numbers per mass of fuel less than 10^{13}. The intermediate or transition regime is between 10^{13} and 10^{14}. Reducing soot in the soot-rich regime yields lower nucleated ice crystal number per mass of fuel. Further reductions of soot in the soot-poor regime may not yield similar reductions of nucleated ice crystal number due to the activation and participation of ultrafine aqueous (i.e., nucleation size mode) particles in the contrail ice nucleation process.
SOURCE: B. Kärcher, 2018, "Formation and Radiative Forcing of Contrail Cirrus," *Nature Communications* 9:1824, https://doi.org/10.1038/s41467-018-04068-0. CC BY-NC-ND 4.0.

Advanced aircraft engine technologies and changes in fuel composition (i.e., SAF or alternative fuels) could reduce the concentrations of nvPM and vPM and the resulting climate impact of persistent contrails and contrail cirrus.

Combustion Technology Effects

Assuming that all of the fuel is completely combusted, the major products of combustion—H_2O, CO_2, O_2, and N_2—are controlled by overall fuel chemistry and fuel-to-air ratio. In other words, the specific combustion technology does not significantly influence these major products. In contrast, minor combustion species, such as CO, NOx, particulates, and volatile organic compounds, are strongly influenced by combustion system architectures. This section provides an overview of these combustion system architectures and their influences on trace species emissions. Further details on these can be found in the references at the end of this chapter.

First, combustion systems can operate in premixed or nonpremixed mode. Notional sketches of each are provided in Figures 2-3 and 2-4.

In a premixed system, the fuel and air are premixed ahead of the combustor. For liquid-fueled combustors encountered in aviation systems, complete premixing requires pre-vaporization of the fuel, which introduces additional operational challenges associated with fuel coking and autoignition. For this reason, aviation combustors typically use architectures where the fuel is introduced at numerous injection points into the combustor, where the large mixing surface area promotes mixing of the fuel and air before combustion. Premixing provides two significant benefits—first, it avoids high-temperature overshoots in regions of stoichiometric combustion and, thus, minimizes NOx emissions. Second, hot fuel is always in the presence of oxidizer and so preferentially reacts to H_2O and CO_2 and produces relatively low soot emissions. As discussed later, in nonpremixed systems, hot fuel in the absence of oxidizer can pyrolize, starting the pathway toward solid, carbonaceous particulate emissions. Premixed systems are the de facto standard in low-NOx, ground-based systems using gaseous fuel. Their use in liquid-fueled systems is more limited, due to operational challenges described above.

Finding: Advanced engine technologies that reduce particulate emissions may play a role in mitigating contrail radiative forcing due to the influence of particulate emissions on contrail dynamics. These technology levers include advanced combustor designs and vent oil management.

The second major combustor classification is the manner of fuel staging—how the fuel is introduced and controlled (in terms of fuel flow rates and injection locations) versus the engine operating condition. While the overall

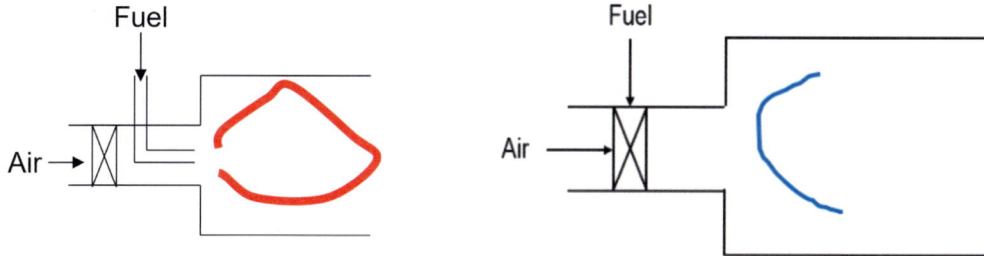

FIGURE 2-3 Notional sketch of nonpremixed (*left*) and premixed (*right*) combustion systems.

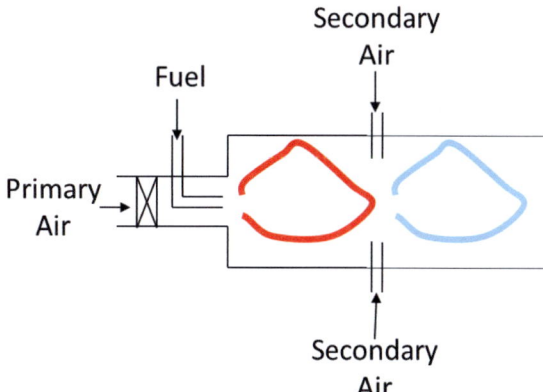

FIGURE 2-4 Notional sketch of a rich-burn, quick-quench, lean-burn (RQL) system, where red line denotes a rich flame and blue line denotes a secondary lean flame.

fuel-to-air ratio is set by the engine cycle (i.e., the desired turbine inlet temperature), the fuel and air introduction into the combustor is managed to enable cooling and emissions control.

A common historical combustor design approach is the RQL combustor technology. An RQL combustor is divided into two main zones. In the primary zone, the combustor is operated fuel rich, with a fraction of the overall air entering the front end of the combustor. The remaining air enters the combustor in the "quench zone" and reacts with the unburned fuel (which, due to its being at a very high temperature, is no longer jet fuel but has decomposed to a synthesis gas blend of H_2 and CO). This fuel-rich zone leads to significant soot production. The majority, but not all, of this soot then reacts with air and is oxidized to CO_2 in the lean zone. The part that does not react results in engine exhaust particulate emissions.

Many aircraft engines in service utilize lean-burn, premixed fuel systems where the primary zone is also operated in fuel-lean combustion. These strategies have demonstrated significantly reduced NOx and soot emissions.

Operating Conditions Effects

Engine emissions vary significantly across the range of engine operating conditions, as power is managed throughout the flight cycle. Lower power (idle and descent) conditions are associated with lower compressor pressure ratio; accordingly, combustion occurs at lower pressure and temperature conditions. Take-off conditions occur on the ground, where the ambient pressure is ~101 kPa; however, most aviation fuel is burned at cruise conditions, which occurs at an altitude where the ambient pressure is ~80 percent lower.

The design of the engine thermodynamic cycle also influences emissions. In a drive for improved thermal efficiency, design trends have increased engine overall compression pressure ratios and turbine inlet temperatures. This results in increased combustor flame temperatures, which is adverse for NOx emissions. Particulate emissions generally increase under conditions where nonpremixed and/or fuel-rich combustion occurs, where atomization and vaporization rates are lower. Both are influenced by pressure, temperature, and fuel composition. In RQL combustion systems, the primary combustion-zone fuel-to-air ratio peaks in rich, high-power conditions, which are also adverse for particulate emissions.

Fuel Composition Effects

Jet fuel composition has a significant influence on particulate emissions from the engine. Jet fuels consist of a range of hydrocarbons that meet the technical and certification requirements for aviation per ASTM standards D1655 and D4054. These requirements on jet fuels are due to their origins from fossil fuels and include a range of specifications around heating value, density, viscosity, and other properties. There is significant interest in synthesizing hydrocarbon jet fuels, known as SAFs, using renewable feedstocks (biomass, etc.). Since the source of SAF feedstocks is from the biosphere, vs petroleum-based conventional jet fuel, they are an attractive approach to reducing the CO_2 impact of aviation.

For SAFs, it is common to differentiate between "drop-in fuels" and "non-drop-in" fuels. A drop-in fuel meets all of the ASTM specifications and certification requirements for commercial jet fuel (Jet A) and does not require the modification of current aircraft, engines, or fueling infrastructure for commercial use. However, non-drop-in fuels do not meet the fuel specifications for commercial use. Accordingly, a non-drop-in fuel formulation would need to be certified to ensure that there is no adverse impact to safety, performance, operability, or durability. Each type of aircraft and engine model would need to be certified for each non-drop-in fuel prior to commercial use. Moreover, separate fueling infrastructure at airports would need to be established to permit the segregation and safe handling, storage, and transportation of these non-drop-in fuels. Conventional jet fuels consist of a wide range of organic compounds. See Box 2-1.

Although aromatics, and particularly polycyclic aromatics, in jet fuel are significant sources of non-volatile particulate matter emissions, there are a number of practical reasons that make it difficult to simply remove them. Aromatic hydrocarbons exhibit bulk properties that enable jet fuels to meet ASTM specifications, where the primary focus is maintaining safety standards. Jet fuels are also used to cool critical hardware components, such as fuel nozzles, so their thermal properties must be sufficient for this purpose. Aromatic compounds have higher density

BOX 2-1
Conventional Jet Fuel Chemistry

Conventional jet fuels consist of a range of hydrocarbons, as shown in Figure 2-1-1. The combustion of aromatic hydrocarbons generates soot particles that permit the formation of ice nuclei in aircraft exhaust plumes. Polycyclic aromatic hydrocarbons (e.g., naphthalenes) are known to be particularly important soot precursors. See Figures 2-1-2 and 2-1-3.

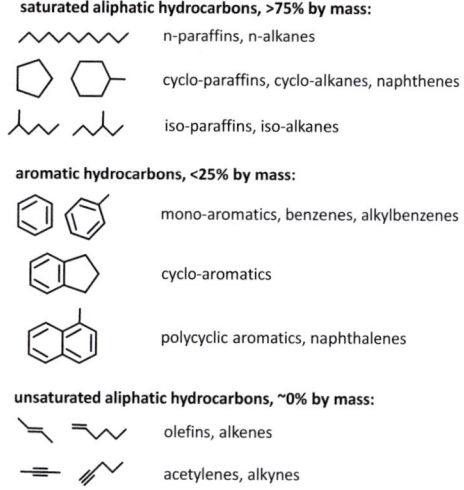

FIGURE 2-1-1 Example hydrocarbon molecules, compound classes, and terminology related to jet fuel composition. The relative contributions of each hydrocarbon class vary across conventional jet fuels and sustainable aviation fuels. Fuel names on each line are synonymous.

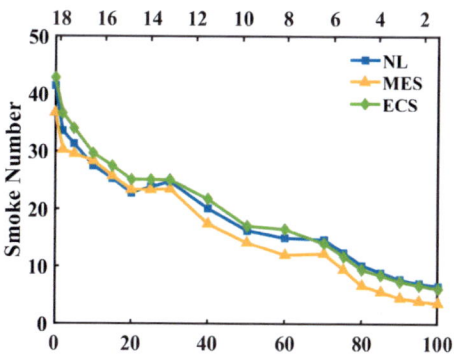

FIGURE 2-1-2 Sooting tendency of HEFA (a class of sustainable aviation fuel derived from hydroprocessed esters and fatty acids) fuel blends showing how decreased aromatic content leads to lower smoke number emissions. A fuel at 0 percent HEFA represents commercial jet fuel. Lines represent the following different engine conditions: no load (NL), main engine start (MES), and environmental control system (ECS).
SOURCE: Produced by T. Lieuwen, I. Gupta, and B. Emerson, based on data from V. Undavalli, J. Hamilton, E. Ubogu, I. Ahmed, and B. Khandelwal, 2022, "Impact of HEFA Fuel Properties on Gaseous Emissions and Smoke Number in a Gas Turbine Engine," *Proceedings of the ASME Turbo Expo 2022: Turbomachinery Technical Conference and Exposition. Volume 3B: Combustion, Fuels, and Emissions*, https://doi.org/10.1115/GT2022-82201.

continued

BOX 2-1 Continued

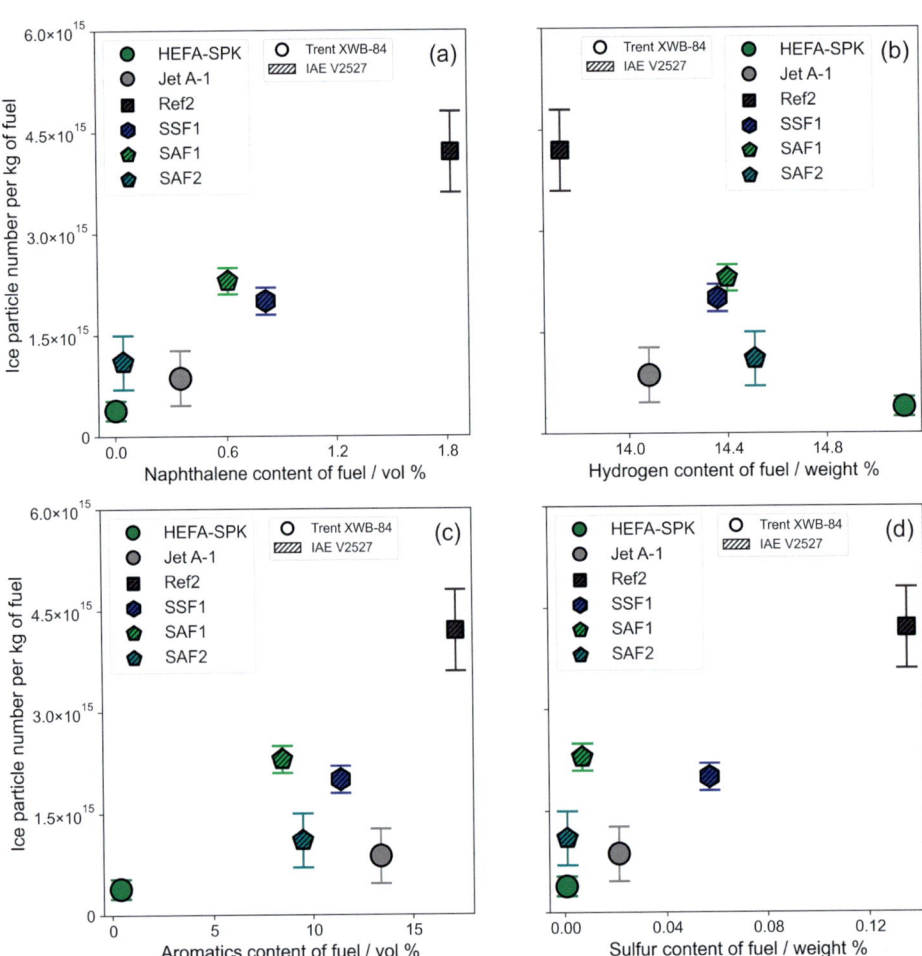

FIGURE 2-1-3 Apparent contrail ice crystal numbers (normalized for engine fuel burn) measured behind rich-burn, quick-quench, lean-burn (RQL)-type engines during recent aircraft campaigns, including ECLIF1, ECLIF2/NDMAX, and ECLIF3. Data are plotted versus fuel parameters for the investigated fuels of co-varying composition: two conventional petroleum-based fuels (Jet A-1 and Ref2), a semi-synthetic jet fuel blend (SSF1), a hydrotreated esters and fatty acids synthetic paraffinic kerosene SAF (HEFA-SPK), and two different HEFA-based sustainable aviation fuel blends (SAF1 and SAF2).
SOURCE: R.S. Märkl, C. Voigt, D. Sauer, et al., 2024, "Powering Aircraft with 100% Sustainable Aviation Fuel Reduces Ice Crystals in Contrails," *Atmospheric Chemistry and Physics* 24(6):3813–3837, https://doi.org/10.5194/acp-24-3813-2024. CC BY-NC-ND 4.0.

and higher energy per unit volume than paraffinic compounds, whereas paraffinic compounds have higher energy per unit mass than aromatic compounds. As a result, fuels that lack aromatic compounds will have considerable challenges meeting the density requirement of ASTM D1655. Highly paraffinic fuels may also have kinematic viscosity and coefficients of specific heat that are outside of the range of historical experience in aviation. Highly paraffinic jet fuels with no aromatics may also exhibit distillation temperature ranges that fall outside of the historical experience in aviation of jet fuels, which would impact aircraft engine operability. The absence of aromatic hydrocarbons in jet fuels also impacts elastomeric seal swelling characteristics and fuel gauging characteristics.

Fuel-bound sulfur molecules chemically react with available oxygen to produce SO_2. A minor fraction of SO_2 will rapidly oxidize into sulfuric acid, which can quickly condense to form sulfur-containing aerosols in the exhaust. These aerosols may serve as nuclei for water condensation and ice crystal growth in linear contrails in ice-supersaturated regions. However, removing sulfur from jet fuels by hydrotreatment also removes oxygenated organics that provide fuel lubricity. Thus, sulfur reduction in fossil jet fuel may lead to lower lubricity, adversely impacting the performance and durability of fuel pumps and valves, requiring fuel additives to mitigate.

SAFs manufactured using the Fischer–Tropsch, hydroprocessed esters and fatty acids (HEFA), and alcohol-to-jet pathways are synthetic paraffinic kerosenes and do not contain aromatic hydrocarbons or sulfur. As a result, the combustion of these synthetic fuels yields lower soot emissions than the combustion of conventional jet fuel, as well as zero SO_2 emissions, substantially reducing the number of cloud condensation nuclei that permit ice crystal and contrail formation. These fuels are classified as non-drop-in fuels, and their commercial use would require modifications to existing aircraft and fueling infrastructure. The current ASTM specifications permit the blending (up to 50 percent for most pathways) of these synthetic paraffinic kerosenes with conventional, fossil-derived jet fuels, such that blended fuel properties satisfy the conventional fuel specifications. SAF blends that meet these ASTM D7566 fuel specifications and historical ranges for the fit-for-purpose properties (such as volumetric energy density, lubricity, kinematic viscosity, surface tension, the coefficient of specific heat, and electrical conductivity) are drop-in compatible with existing aircraft and fueling infrastructure. The ASTM D7566 requirements also include a minimum limit on the aromatic hydrocarbon concentration by volume of 8 percent, whereas conventional jet fuels do not have a minimum limit on aromatic hydrocarbon concentration.

There are several federally supported programs in place to both develop SAF, as well as to evaluate influences on engine emissions, operability, and stability. These include programs in the Department of Energy (DOE), the Department of Defense, the Federal Aviation Administration (FAA), and NASA (Colket et al. 2017; DOE 2024). Moreover, FAA supports the Commercial Alternative Aviation Fuels Initiative (CAAFI), which promotes alternative fuels that meet equivalent safety requirements of conventional jet fuels.[2] Additionally, CAAFI's Fuels Certification and Qualification and Research and Development teams are tasked with evaluating new sustainable aviation fuels per a standard approval process developed by ASTM and evaluating novel sustainable aviation fuel production pathways, respectively. NASA partners with FAA, universities, airframers, engine manufacturers, and fuel companies to evaluate the impact of alternative jet fuels on aircraft emissions and contrail properties.

Finally, there is a broader palate of potential future fuels that are not drop-in and are not currently used in aviation. Hydrogen combustion leads to only three major combustion products—H_2O, O_2, and nitrogen oxides. Hydrogen combustion does produce elevated water vapor emissions, but no direct particulate production from the fuel. However, contrails can still form due to nucleation of ambient environmental particulate matter. Particulates can also be produced from venting of the lubrication oil.

Also being considered as future fuels are other hydrocarbons, such as methane. There is significant experience in the ground power community with methane-air combustion. In this case, the lower carbon-to-hydrogen ratio and easier ability to premix both lead to reduced particulates. It is uncertain whether the contrail formation will be dominated by ambient particulate matter or engine-generated particulate matter for such.

---

[2] See the Commercial Aviation Alternative Fuels Initiative website at https://www.caafi.org/about, accessed December 1, 2024.

LUBRICATION OIL VENT EFFECTS

Aircraft gas turbines require specially formulated oil to lubricate, cool, and provide corrosion resistance to the compressors, bearings, gears, and other turbomachinery components. The oil is recirculated through the engine systems, including a recycling loop that is designed to minimize oil consumption. During typical engine operation, the action of the rotating machinery in the lubrication system environment causes some small amount of oil to be mixed with air to form a mist. Typically, air/oil separator components recover the majority of the misted oil, but some small amount of the air/oil mixture is vented overboard to the atmosphere. This vented oil mist can have significant impacts on contrail formation, depending on the location of the breather vent, which impacts the number and size distribution of the emitted semi-volatile oil particles. Common design choices include a vent on the outside of the engine nacelle (e.g., as in Figure 2-5), into the bypass air, or at the tail cone of the engine core. Given the close proximity of the tail cone vent to the hot core exhaust flow, any condensed oil droplets likely evaporate as they mix with the hot combustion products—either recondensing on co-emitted soot (thermodynamically preferable) or forming new nucleation-mode particles via gas-to-particle conversion (less thermodynamically preferable). In this case, the amount of soot emissions may have a significant impact on the resulting distribution of aerosol-size oil droplets. Gas turbines operating in the soot-poor or no-soot (e.g., hydrogen) regimes are more likely to experience elevated oil-based new particle formation compared to gas turbines that operate in the soot-rich regime with ample soot surface area to act as a condensation sink. For breather vent locations in the engine bypass flow and outside of the nacelle, some or all of the vented, condensed oil droplets may survive as a larger, externally mixed, aerosol-size distribution mode, while the oil vapor condenses onto the oil particles, as well as onto the pre-existing soot and nucleation mode particles. However, the number of droplets in such a larger, externally mixed-size mode is unlikely to be significant for contrail formation, compared to typically much larger numbers of co-emitted nucleation-mode and soot-mode particles.

The oil contributions to particles larger than ~50 nm in diameter are readily apparent from aerosol mass spectrometer (AMS) measurements during emissions ground tests conducted over the past two decades (Timko et al. 2010; Yu et al. 2010, 2012, 2019). These studies, and also more recent thermal desorption–gas chromatography–mass spectrometry work, show substantial aerosol mass contributions from nearly intact forms of lubrication oil, indicating no apparent thermal degradation or chemical transformation of the oil (Fushimi et al. 2019; Yu et al.

FIGURE 2-5 Example of an oil vent diversion system employed during the 2022 NASA Boeing ecoDemonstrator Emissions Ground Test connected to the Rolls-Royce Trent 800 oil vent port on the bottom of the nacelle. The white oil vapor plume is clearly visible at the bottom left side of the image.
SOURCE: Copyright © Boeing.

2010). Generally, a larger oil contribution to the particulate mass is observed at high engine thrust than at idle. An important limitation of these studies is the inability of current online and filter-based offline measurements to isolate and quantify the chemical composition of the smaller nucleation-particle-size mode. This is because the AMS is unable to transmit these particles through its inlet and because mass compositional techniques measuring the total aerosol mass will be weighted toward the largest, most massive, particles. While the nucleation-size mode contributes large numbers of particles (hence the potential importance of this mode for contrails in the soot-poor regime), the mass contribution from the nucleation-size mode is negligible. Previous ground test measurements of the vented oil contribution to engine particle emissions are limited and non-existent at cruise conditions relevant for contrail formation. More measurements are needed with particular emphasis on soot-poor engines.

Finding: Aircraft engine particulate emissions are strongly influenced by fuel composition, engine operating conditions, lubrication oil venting, and combustion system technology. Measuring the size-resolved chemical composition of these nucleation-mode particles is a major challenge, with gaps in predictive capabilities.

Finding: The largest impact of alternative fuels on contrails is likely to be through changes in the particulate content of aircraft exhaust.

Finding: For no- or low-soot-emitting engine technologies, and for fuel compositions with inherently low particulate formation tendencies (including SAFs), lubrication oil systems play an important role in the engine particulate emissions, which can serve as condensation nuclei for contrail ice crystals formed behind soot-poor and non-soot-emitting engines.

Finding: There are a number of emission sources that contribute to nucleation-mode particles, including fuel sulfur, unburned hydrocarbons, and lubrication oil. Developing the ability to identify the composition and roles of these contributors is needed to predict how changes in fuel composition and engine technology will influence engine particle emissions relevant for contrail formation. More measurements of the ice-nucleating properties of contrail-processed soot, and of ice nucleation concentrations in the upper troposphere, are needed to better constrain this issue.

Recommendation: NASA, in coordination with the Federal Aviation Administration, the Department of Energy, the Department of Defense, other relevant federal agencies, and the private sector, should support development of low-particle-emitting combustion technologies, as well as sustainable aviation fuels with inherently low particulate-formation tendencies.

Recommendation: NASA, in coordination with the Federal Aviation Administration, the Department of Energy, and the Department of Defense, should support laboratory and engine research studies to improve the understanding of how fuel composition, combustor technology, and engine operating conditions impact particulate emissions (volatile and non-volatile) and contrail properties.

PARTICLE IMPACTS ON CONTRAIL FORMATION AND DOWNSTREAM AEROSOL-CLOUD INTERACTIONS

Aircraft particle emissions are thought to impact clouds and climate via two pathways: first, through the formation and persistence of line-shape contrails and diffuse contrail cirrus (together referred to as "aviation-induced cloudiness"), and second, through contributing cloud condensation nuclei and ice nucleating particles to naturally occurring clouds (referred to as "aerosol-cloud interactions").

First, particulates can nucleate water droplets that homogeneously freeze into contrail ice crystals if the conditions satisfy the Schmidt–Appleman criterion. These contrail ice crystals will persist if the ambient atmosphere is supersaturated with respect to ice, as discussed in Chapter 1. Given the high-water-vapor supersaturation with respect to liquid in the initial contrail-forming plume, it is expected that almost all the particles will uptake water if they are above a certain critical size (somewhere in the range 10–40 nm depending on the local supersaturation in the highly heterogeneous plume), which is largely independent of particle composition. This has been verified with in situ observations of contrail ice crystal numbers compared to particulates; however, the flight studies to date have exclusively considered emissions from jet fuels with non-negligible amounts of fuel sulfur (either by burning sulfur-containing Jet A or nominally zero-sulfur alternative fuels with trace amounts of sulfur contamination introduced during post-refinery fuel transport and handling). Open research questions remain about contrail ice formation for a truly zero-sulfur fuel for aircraft engines operating in both the soot-rich and soot-poor contrail regimes.

Second, if contrails do not form immediately behind an aircraft, or once the contrail ice crystals sublimate, the soot particles have been hypothesized to act as ice nuclei for subsequent naturally occurring cirrus clouds that might form, although one recent laboratory experiment with aircraft soot does not support this (Testa et al. 2024). The properties of the emitted particles evolve in the atmosphere as the result of different processes. For example, particles that become part of a contrail may be changed (due to condensation and freezing) and redistributed in the atmosphere (due to sedimentation). Particulates also "age" in the atmosphere due to heterogeneous chemistry with reactive gases (e.g., NOx, ozone) and condensation and coagulation with both sulfate particles and organics. In addition to the formation of ice-nucleating particles, secondary gas-to-particle formation of hygroscopic sulfate and nitrate aerosols may contribute cloud-condensation nuclei to warm clouds; however, the contribution of aviation particles is likely to be much smaller than that from terrestrial sources. More measurements of the ice-nucleating properties of contrail processed soot and of ice-nucleating particle concentrations in the upper troposphere are needed to better constrain this issue.

Much of the uncertainty in the effects of aircraft emissions on ice particle concentrations for natural cirrus clouds stems from uncertainty in the ice-nucleating particles and their properties in the background atmosphere. Liquid aerosol particles (sulfate or liquid organic particles) typically require very high supersaturations to form ice via homogeneous freezing, whereas many solid or semi-solid particles can form ice at lower ice supersaturation but have very low number concentrations. The presence, or lack thereof, of background heterogeneous ice nucleation in the upper troposphere has a major control over contrail ice crystal concentration and, therefore, presents a major uncertainty in contrail radiative forcing, especially in the low-soot regime. There is low confidence in the calculation of the effect of aircraft soot on naturally occurring cirrus. It has only been assessed in a few studies (summarized in Lee et al. 2021). Moreover, it is difficult to attribute observed atmospheric soot in the upper troposphere directly to aviation operations, due to the significant amount of other non-volatile particles that may be found at flight levels due to convective transport of terrestrial and volcanic dust, as well as soot from other sources.

These aerosol effects are noted as "Aerosol Cloud Interactions" in Figure 1-3 in Chapter 1. Lee et al. (2021) did not put an estimate or an error range on these effects due to the diversity of approaches taken in the few studies that have investigated aerosol-cloud interactions from aviation particulates. Most estimates indicate a moderate cooling from sulfate and/or soot, with one sensitivity study (Zhou and Penner 2014) indicating a large warming or cooling depending on the treatment of background aerosols. Another study (Gettelman and Chen 2013) noted that aviation sulfate aerosols could alter liquid clouds and create a significant cooling effect if aerosols get to lower altitudes. Further work is needed to constrain and evaluate these model estimates.

> Finding: Aviation-induced cloudiness from aviation particulates outside of a contrail is highly uncertain and needs to be better constrained with further development of modeling studies tied to more observations of aviation and background aerosols, in particular, heterogeneous ice nuclei at flight level.

NASA has a long history of collecting airborne aerosol information in the upper troposphere, including from contrails specifically. Airborne science campaigns are exploring the impacts of alternative fuels and new combustor technology on aviation particulate emissions (Figure 2-6). For example, the NASA-led ACCESS flight test series in 2013–2014 demonstrated cruise altitude particle emissions reductions from burning a 50 percent SAF blend in the CFM56-2C engines of the NASA DC-8 (Moore et al. 2017). The ECLIF-2/ND-MAX flight test in 2018 that examined the soot-particle emissions from specially sourced SAF blends with varying levels of fuel aromatic and naphthalenic contents showed similar reductions in cruise soot-particle emissions for IAE V2500 series engines and linked these soot-particle reductions to contrail ice crystal reductions (Voigt et al. 2021). More recent tests

FIGURE 2-6 NASA has conducted numerous flight campaigns to measure various aspects of contrails, including flights of the DC-8 research aircraft shown here, using synthetic fuels.
SOURCES: (a) Courtesy of NASA/Andy Barry; (b) Courtesy of NASA/Lori Losey; (c) Courtesy of NASA/Eddie Winstead.

have focused on contrails and cruise emissions for 100 percent SAF in RQL engines (e.g., DLR ECLIF-3; Märkl et al. 2024) and lean-burn engines (e.g., NASA Boeing ecoDemonstrator, DLR Airbus VOLCAN experiments). Large databases[3] and statistics from many flights are needed due to the large variability of observed aviation particulates and contrail microphysics. Despite this track record, many of the current contrail and aerosol particle sensors being deployed by NASA and other research institutions for contrail studies are decades old, which places these capabilities at risk for future flight test projects. There is a need to update and maintain these aging sensor capabilities as well as to stimulate a robust commercial and academic pipeline of sensor development targeting high-altitude aerosol and contrail (i.e., small ice) measurements.

Recommendation: NASA should continue to collect in-flight observational data of contrails and cruise emissions (CO_2, NOx, and ice-nucleating particles) from aviation that advance the understanding of the factors that influence contrail properties.

Chapter 3 discusses methods of gathering the necessary data to inform the models.

REFERENCES

Ahrens, D., Y. Méry, A. Guénard, and R.C. Miake-Lye. 2023. "A New Approach to Estimate Particulate Matter Emissions from Ground Certification Data: The nvPM Mission Emissions Estimation Methodology." *ASME Journal of Engineering for Gas Turbines and Power* 145(3):031019. https://doi.org/10.1115/1.4055477.

Colket, M., J. Heyne, M. Rumizen, M. Gupta, T. Edwards, W.M. Roquemore, G. Andac, et al. 2017. "Overview of the National Jet Fuels Combustion Program." *AIAA Journal* 55(4):1087–1104.

DOE (Department of Energy). 2024. *Pathways to Commercial Liftoff: Sustainable Aviation Fuel*. https://liftoff.energy.gov/sustainable-aviation-fuel-2.

Durdina, L., B.T. Brem, A. Setyan, F. Siegerist, T. Rindlisbacher, and J. Wang. 2017. "Assessment of Particle Pollution from Jetliners: From Smoke Visibility to Nanoparticle Counting." *Environmental Science & Technology* 51(6):3534–3541. https://doi.org/10.1021/acs.est.6b05801.

Fushimi, A., K. Saitoh, Y. Fujitani, and N. Takegawa. 2019. "Identification of Jet Lubrication Oil as a Major Component of Aircraft Exhaust Nanoparticles." *Atmospheric Chemistry and Physics* 19(9):6389–6399.

Gettelman, A., and C-C. Chen. 2013. "The Climate Impact of Aviation Aerosols." *Geophysical Research Letters* 40(11):2785–2789. https://doi.org/10.1002/grl.50520.

Kärcher, B. 2018. "Formation and Radiative Forcing of Contrail Cirrus." *Nature Communications* 9(1):1824.

Lee, D.S., D.W. Fahey, A. Skowron, M.R. Allen, U. Burkhardt, Q. Chen, S.J. Doherty, et al. 2021. "The Contribution of Global Aviation to Anthropogenic Climate Forcing for 2000 to 2018." *Atmospheric Environment* 244:117834.

Märkl, R.S., C. Voigt, D. Sauer, R.K. Disch1, S. Kaufmann, T. Harlaß, V. Hahn, et al. 2024. "Powering Aircraft with 100% Sustainable Aviation Fuel Reduces Ice Crystals in Contrails." *Atmospheric Chemistry and Physics* 24:3813–3837.

Moore, R.H., K.L. Thornhill, B. Weinzierl, D. Sauer, E. D'Ascoli, J. Kim, M. Lichtenstern, et al. 2017. "Biofuel Blending Reduces Particle Emissions from Aircraft Engines at Cruise Conditions." *Nature* 543:411–415. https://doi.org/10.1038/nature21420.

Peck, J., O.O. Oluwole, H.W. Wong, and R.C. Miake-Lye. 2013. "An Algorithm to Estimate Aircraft Cruise Black Carbon Emissions for Use in Developing a Cruise Emissions Inventory." *Journal of the Air Waste and Management Association* 63(3):367–375. https://doi.org/10.1080/10962247.2012.751467.

Petzold, A., A. Döpelheuer, C.A. Brock, and F. Schröder. 1999. "In Situ Observations and Model Calculations of Black Carbon Emission by Aircraft at Cruise Altitude." *Journal of Geophysical Research* 104(D18):22171–22181. https://doi.org/10.1029/1999JD900460.

Stettler, M., A.M. Boies, A. Petzold, and S. Barrett. 2013. "Global Civil Aviation Black Carbon Emissions." *Environmental Science & Technology* 47(18):10397–10404. https://doi.org/10.1021/es401356.

Testa, B., L. Durdina, J. Edebeli, C. Spirig, and Z..A. Kanji. 2024. "Contrail Processed Aviation Soot Aerosol Are Poor Ice Nucleating Particles at Cirrus Temperatures." *Atmospheric Chemistry and Physics* 24(18):10409–10424. https://doi.org/10.5194/acp-24-10409-2024.

---

[3] See, for example, NASA, "Aeronautics Field Projects," Aeronautics Research Mission Directorate, last modified April 5, 2022, https://science.larc.nasa.gov/aero-fp.

Timko, M.T., T.B. Onasch, M.J. Northway, J.T. Jayne, M.R. Canagaratna, S.C. Herndon, E.C. Wood, R.C. Miake-Lye, and W.B. Knighton. 2010. "Gas Turbine Engine Emissions—Part II: Chemical Properties of Particulate Matter." *Journal of Engineering for Gas Turbines and Power* 132(6):061505.

Voigt, C., J. Kleine, D. Sauer, R.H. Moore, T. Bräuer, P. Le Clercq, S. Kaufmann, et al. 2021. "Cleaner Burning Aviation Fuels Can Reduce Contrail Cloudiness." *Communications Earth and Environment* 2(114).

Yu, Z., D.S. Liscinsky, E.L. Winstead, B.S. True, M.T. Timko, A. Bhargava, S.C. Herndon, R.C. Miake-Lye, and B.E. Anderson. 2010. "Characterization of Lubrication Oil Emissions from Aircraft Engines." *Environmental Science & Technology* 44(24):9530–9534.

Yu, Z., S.C. Herndon, L.D. Ziemba, M.T. Timko, D.S. Liscinsky, B.E. Anderson, and R.C. Miake-Lye. 2012. "Identification of Lubrication Oil in the Particulate Matter Emissions from Engine Exhaust of In-Service Commercial Aircraft." *Environmental Science & Technology* 46(17):9630–9637.

Yu, Z., M.T. Timko, S.C. Herndon, R.C. Miake-Lye, A.J. Beyersdorf, L.D. Ziemba, E.L. Winstead, and B.E. Anderson. 2019. "Mode-Specific, Semi-Volatile Chemical Composition of Particulate Matter Emissions from a Commercial Gas Turbine Aircraft Engine." *Atmospheric Environment* 218:116974.

Zhou, C., and J.E. Penner. 2014. "Aircraft Soot Indirect Effect on Large-Scale Cirrus Clouds: Is the Indirect Forcing by Aircraft Soot Positive or Negative?" *Journal of Geophysical Research: Atmospheres* 119(19):11303–11320. https://doi.org/10.1002/2014JD021914.

3

Atmospheric Measurements

Atmospheric measurements of meteorology, naturally occurring aerosols, and contrail properties are necessary to validate contrail prediction models and to improve scientific understanding of where contrails form and persist. Such measurements may be from ground- and space-based remote sensors, as well as airborne in situ platforms such as aircraft and balloons. This chapter describes measurement needs for identifying ice-supersaturated regions (ISSRs) in the atmosphere and discusses research to characterize the naturally occurring atmospheric aerosol background of the upper troposphere.

Model predictions of atmospheric ice supersaturation and contrail formation/persistence can be improved by ingesting cruise altitude meteorological observations, of which the measurement of water vapor is the most critical. There is a need to develop new in situ water vapor sensors with improved reliability and accuracy to fly autonomously and downlink data from the commercial aircraft fleet in near real time. These in situ aircraft measurements are the most relevant and cost-effective approach to improving numerical weather prediction models, as demonstrated by current airborne observation networks complemented by weather balloons and satellite remote sensing. Satellite temperature and water vapor soundings (i.e., vertical profiles) hold promise for global model improvement, while space- and ground-based imagers aid in contrail detection and monitoring that could be enhanced by collocation with aircraft flight track information. Methods for assimilating remote sensing observations into models are an important area of active research. Camera imagery (ground, airborne, and satellite) captures the essence of contrails as human fingerprints on the Earth system and presents opportunities for public outreach and education as well as citizen-science activities to develop robust data sets for testing model predictions and remote sensing retrievals. Finally, this chapter highlights the need for improved characterization of upper tropospheric background aerosols to better understand the role of these particles in contrail formation for "soot-poor" engines as well as to quantify the aviation emissions impacts on aerosol-cloud interactions within naturally occurring cirrus clouds.

MEASUREMENTS OF ATMOSPHERIC STATE PARAMETERS

As discussed in Chapter 1, contrails form when the cooling aircraft engine exhaust plume becomes supersaturated with respect to liquid water (i.e., the Schmidt–Appleman criterion; Appleman 1953; Schmidt 1963; Schumann 1996), which depends on the atmospheric state parameters (temperature, pressure, winds, and humidity), as well as engine and fuel characteristics. The contrail ice crystals further persist and grow into a contrail cirrus cloud as long as the surrounding atmosphere is supersaturated with respect to ice; otherwise, the contrail sublimates. Since only persistent contrail cirrus are climatically relevant, it is critical to be able to forecast and diagnose these ISSRs of the atmosphere, which are thought to be horizontally vast (~hundreds of kilometers) but vertically shallow (~hundreds of meters) (Sausen et al. 2024; Spichtinger and Leschner 2016). However, this understanding is limited by the generally coarse vertical resolution of current satellite remote sensors (~2–3 km) and a dearth of in situ observations of temperature and humidity, particularly away from land-based radiosonde networks.

Finding: A large uncertainty in characterizing the location and extent of persistent contrail formation is the ability to observe and/or predict ISSRs. Ice supersaturation depends on the upper tropospheric temperature and humidity.

Recommendation: NASA should support research and observational studies to improve the understanding of the extent and frequency of ice-supersaturated regions (ISSRs) and the level of skill in simulating ISSRs and contrails.

Aircraft-, balloon-, and satellite-based observations are critical, complementary data for constraining numerical weather prediction model forecasts and nowcasts that underpin the ability to diagnose ISSRs as well as broader societal weather concerns. In the United States, observational data are ingested into the National Weather Service (NWS) Meteorological Assimilation Data Ingest System (MADIS), including the following:

- Balloon-borne radiosondes[1]
- Aircraft-based observations[2]
- Satellite soundings[3]

Radiosondes measuring temperature and humidity profiles are released one to four times per day and have good coverage over North America and Europe (typically <500 km horizontal separation), but they are sparse or non-existent over much of the globe. Radiosondes are costly (about $300 each) and also raise concerns about their ecological impacts. Aircraft-based operational soundings overcome these limitations through deployment of reusable sensors that bring the cost of a sounding down significantly (Marshall 2024), but commercial sensors certified for commercial aircraft currently lack adequate accuracy and reliability to measure humidity at cruise altitudes. Meanwhile, satellite soundings possess global coverage, but their limited vertical resolution is not a replacement for the high-quality information afforded by in situ atmospheric sensors.

Finding: Accurate, high spatial (particularly vertical) resolution measurements of humidity and temperature are needed to constrain model forecasts/nowcasts of contrail-forming conditions (i.e., Schmidt–Appleman criterion) and cruise-level ISSRs.

Aircraft are particularly attractive platforms for capturing cruise-level temperature and humidity data and studying contrail formation. There are a small number of research aircraft worldwide that are equipped to comprehensively study the atmospheric composition and dynamics, and these aircraft routinely conduct process-level studies and field campaigns related to contrail formation and ISSRs. Large research aircraft such as the recently retired NASA DC-8 and its much anticipated 777-200ER replacement (expected 2026 entry into service) can host a large number of in situ and remote sensing instrument groups. Recent large aircraft experiments such as the 2023 NASA Boeing ecoDemonstrator and 2018 NASA DLR ECLIF-2/ND-MAX campaigns included broad industry, academia, and international participation. Smaller research aircraft such as the NASA Gulfstream III and V aircraft,

---

[1] National Oceanic and Atmospheric Administration (NOAA), "Radiosonde Dataset," National Centers for Environmental Prediction, https://madis.ncep.noaa.gov/madis_raob.shtml, accessed December 1, 2024.

[2] NOAA, "Aircraft Based Observation (ABO) Dataset," National Centers for Environmental Prediction, https://madis.ncep.noaa.gov/madis_acars.shtml, accessed December 1, 2024.

[3] NOAA, "Satellite Sounding Dataset," National Centers for Environmental Prediction, https://madis.ncep.noaa.gov/madis_satsnd.shtml, accessed December 1, 2024.

the DLR Falcon and HALO G-V, and the NCAR G-V are also suitable for targeted studies employing a more limited number of instruments and/or investigators. Some recent relevant small aircraft flight campaigns over the past decade include 2024 CODEX, 2021–2023 DLR VOLCAN, 2021 DLR ECLIF-3, 2021 DLR CIRRUS-HL, 2013-2014 NASA DLR ACCESS, and 2014 DLR ML-CIRRUS.

While such studies are invaluable for advancing the scientific understanding of these phenomena and case study validation and verification, they are too infrequent and spatially sparse to be relied on for operational weather forecast data assimilation. Commercial aircraft, on the other hand, are much more relevant to forecast model ingest and assimilation given their large spatiotemporal coverage at cruise altitudes. These aircraft are also not designed to readily fly complex, state-of-the-art scientific instrumentation or bulky inlets. Instruments must be designed to be easily integrated into existing airframes and avionics, which can be a limiting factor in the number of available meteorological observations. Fortunately, temperature, pressure, and wind sensors are also important flight parameters and are readily available on all commercial aircraft. Water vapor sensors are currently much more limited both in terms of instrument accuracy and calibration stability as well as deployment across the global commercial fleet.

Meteorological Observations from Commercial Aircraft

Commercial aircraft have long provided meteorological soundings of temperature as well as wind speed and wind direction during ascent and descent through the World Meteorological Organization (WMO) Aircraft Meteorological Data Relay (AMDAR[4]) program. In the United States, the National Oceanic and Atmospheric Administration's (NOAA's) Aircraft-Based Observation Program contributes to AMDAR and MADIS through the Aircraft Communications, Addressing, and Reporting System (ACARS). Data are purchased contractually from Collins Aerospace as the ACARS communications provider who owns the air-to-ground very-high-frequency (VHF) communications infrastructure and has subcontracting relationships with almost all of the U.S. commercial airlines. The contract was held by the Federal Aviation Administration (FAA) from the late 1990s until 2016 and then transitioned to the NWS in 2017. Currently, the United States procures and provides about 3 million AMDAR soundings per year worldwide covering 3,500 aircraft in the U.S. commercial fleet. A subset of United Parcel Service and Southwest aircraft (approximately 135 in number) additionally provide water vapor soundings through the Water Vapor Sensing System (WVSS) program. Many of the WVSS program aircraft are older and in line for retirement in the near future, which motivates the need for newer, replacement sensors that can be certified for installation on the current and future commercial aviation fleet. Aircraft profile data are collected at shorter intervals for the landing and take-off cycle (6, 20, and 60 seconds for take-off, climb, and approach segments, respectively). En route cruise altitude data are collected at 3-min intervals (roughly 45 km distance assuming a nominal airspeed of 900 km/hr), and it is an open question whether this measurement frequency is sufficient to resolve ISSRs. The challenge of data frequency is even more acute for satellite communications pathways such the Automatic Dependent Surveillance-Contract datalink system currently providing temperature and winds data with 14 min (i.e., 210 km distance) resolution. The longer en route data intervals are not due to instrument response time and are driven by the need to balance data volume with the communications costs and throughput limitations associated with the VHF and satellite downlink.

While the existing air-to-ground communications and NWS ingest pathways are suitable for connecting aircraft temperature, pressure, and humidity measurements to weather forecasting objectives, there are opportunities to improve and optimize the system specifically to inform ISSR and contrail prediction. First, research is needed to understand the necessary measurement spatiotemporal resolution to resolve vertically thin but horizontally large ice-supersaturated layers, which involves both the number of aircraft sensors deployed across the fleet and their reporting intervals. Second, incorporating meteorological information into new and existing communications technologies could facilitate significant cost and efficiency savings. A particularly promising approach would be to leverage broadband communications pathways currently available for in-flight Internet connectivity

---

[4] World Meteorological Organization, "The WMO AMDAR Observing System," https://community.wmo.int/en/activity-areas/aircraft-based-observations/amdar, accessed December 1, 2024.

in the passenger cabin to increase the information available from the aircraft as a comprehensive atmospheric sensor; however, attendant cybersecurity concerns would need to be considered before connecting aircraft data systems with the in-cabin network. Another approach would be to use the existing weather element placeholder for water vapor to broadcast this information alongside temperature and wind data via the Automatic Dependent Surveillance-Broadcast (ADS-B) system.

In Situ Water Vapor Sensors on Commercial Aircraft

Instrumenting commercial aircraft with a minimum complement of in situ sensors is a promising method for increasing the observational coverage of ISSRs. As discussed above, robust methods for measuring static temperature, static pressure, and winds are commonplace across the fleet, but a key gap is a similarly robust water vapor sensor that has sufficient sensitivity and accuracy to capture the low mixing ratios of the upper troposphere. Figure 3-1 shows the water vapor sensor measurement ranges necessary for the calculating Schmidt–Appleman criterion and ice supersaturation prerequisites for contrail formation and persistence as well as some examples of current sensors that would cover this range. For typical atmospheric temperatures at cruise altitudes (200–300 hPa pressure levels), an instrument with 1–2 ppm accuracy down to 20–30 parts per million by volume (ppmv) is a reasonable goal that would allow for accurate calculations of relative humidity with respect to ice (RHi) near 100 percent at

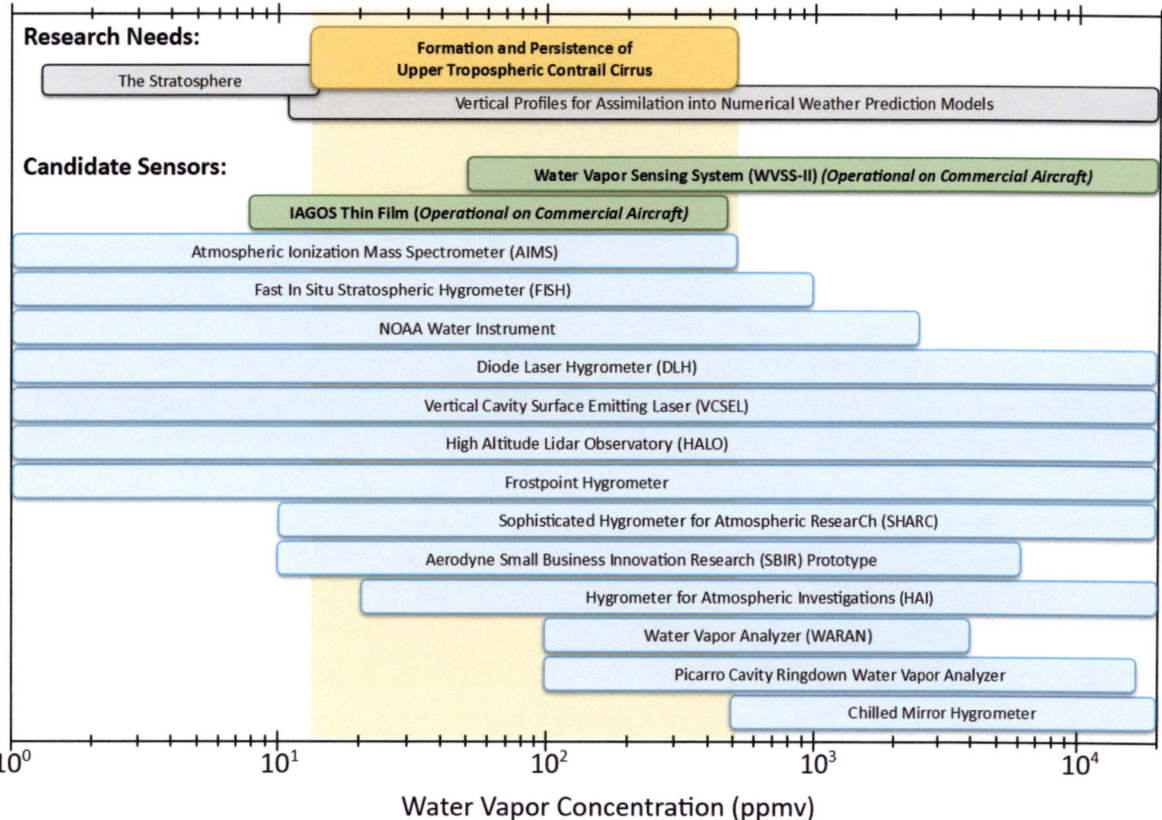

FIGURE 3-1 Measurement ranges of candidate water vapor sensors for commercial aircraft deployment and research topic needs. Sensors that are currently operational on commercial aircraft are shaded green, while research-grade and in-development sensors are shaded blue. None of these sensors are currently adequate for widespread autonomous deployment on commercial aircraft for studying ice-supersaturated regions.
NOTE: IAGOS, In-service Aircraft for a Global Observing System.

temperatures as cold as −68°C. While an upper measurement range of only several hundred parts per million by volume is required for studying contrail formation and persistence, it would be ideal for in situ sensors to also be able to measure near-surface humidity profiles relevant for numerical weather prediction (exceeding 10^4 ppmv) that are relevant for the convective weather and broader aviation weather concerns.

Most of the sensors listed in Figure 3-1 and Table 3-1 are research-grade instruments, and only two instruments are currently operationally deployed on commercial aircraft and reporting cruise altitude measurements: (1) the FLYHT (formerly SpectraSensors) Water Vapor Sensing System (WVSS-II), which operates via tunable diode laser (TDL) spectroscopy; and (2) the In-Service Aircraft for a Global Observing System (IAGOS) thin film capacitive moisture sensor. While the current generation WVSS-II is generally better suited for lower-altitude (i.e., higher

TABLE 3-1 Instrument Specifications and Selected References for Candidate Water Vapor Instruments for Measuring Atmospheric Ice Supersaturation from Commercial and Research Aircraft Shown in Figure 3-1

Instrument	Technique	Installation	Range (ppm)	Resolution(s)	Uncertainty	References
Water Vapor Sensing System (WVSS-II), *Currently Operational on Commercial Aircraft*	Tunable Diode Laser (TDL) Absorption Spectrometry	Inlet	50–60000	0.25–2	5% or 50 ppm	Ford 2011; Vance et al. 2015
IAGOS Thin Film, *Currently Operational on Commercial Aircraft*	Thin Film Capacitive Sensor	External	10–500	60–180	5–10% in relative humidity	Helten et al. 1998; IAGOS 2024
Atmospheric Ionization Mass Spectrometer (AIMS)	Mass Spectrometry	Inlet	1–500	0.25	7–15%	Kaufmann et al. 2016; Thornberry et al. 2013
Fast In Situ Stratospheric Hygrometer (FISH)	Lyman-α Fluorescence	Inlet	1–1000	1	6% or 0.4 ppm	Afchine et al. 2018; Meyer et al. 2015; Schiller et al. 2009
NOAA Water Instrument	TDL Absorption Spectrometry	Inlet	0.5–2500	1	5% or 0.23 ppm	Thornberry et al. 2015
Diode Laser Hygrometer (DLH)	TDL Absorption Spectrometry	Open Path	0.5–40000	0.05	5% or 0.1 ppmv	Diskin et al. 2002
Vertical Cavity Surface Emitting Laser (VCSEL)	TDL Absorption Spectrometry	Open Path	0.1–40000	0.04	5%	Zondlo et al. 2010
High Altitude Lidar Observatory (HALO)	Differential Absorption Lidar (DIAL)	Open Path	1–25000	Variable 5–60 s (horiz.), 250 m (vertical)	Depends on resolution, 5–15%	Carroll et al. 2022
Frostpoint Hygrometer	Chilled Mirror Hygrometer	External	0.8–25000	10–20	0.2 K frostpoint or 10%	Stuefer and Gordon 2018
Sophisticated Hygrometer for Atmospheric ResearCh (SHARC)	TDL Absorption Spectrometry	Inlet	10–50000	1	5% or 1 ppm	Kaufmann et al. 2018

continued

TABLE 3-1 Continued

Instrument	Technique	Installation	Range (ppm)	Resolution(s)	Uncertainty	References
Aerodyne Small Business Innovation Research (SBIR) Prototype	TDL Absorption Spectrometry	Inlet	10–6000	1	0.11 ppm	https://techport.nasa.gov/projects/154493
Hygrometer for Atmospheric Investigations (HAI)	TDL Absorption Spectrometry	Inlet	20–40000	1	4.3% or 3 ppm	Afchine et al. 2018; Buchholz et al. 2017
Water vapoR ANalyzer (WARAN)	TDL Absorption Spectrometry	Inlet	100–40000	0.4	5% or 50 ppm	Afchine et al. 2018
Picarro H2O Analyzer	Cavity Ringdown Spectrometry	Inlet	100–30000	2.8	5% or 100 ppm	Karion et al. 2013; Picarro n.d.
Chilled Mirror Hygrometer	Chilled Mirror Hygrometer	External	500–11000	50–150 (moist-to-dry transition lag times)	0.7 K dewpoint (~5 ppm at 11 km assuming a standard atmosphere)	Buck Research Instruments 2024; PST 2024; Vance et al. 2015

water vapor concentration) measurements, efforts are under way to extend the measurement range of the WVSS-II and to develop new TDL-based sensors that meet the 20–30 ppmv lower sensitivity requirement. Another key challenge is ensuring calibration stability and minimizing sensor-to-sensor variability that has been observed both in flight and in the laboratory. The Department of Energy's (DOE's) Advanced Research Projects Agency–Energy (ARPA-E) has funded a number of promising technology development projects through its Predictive Real-Time Emissions Technologies Reducing Aircraft Induced Lines in the Sky[5] (PRE-TRAILS) program, and NASA and DOE have funded water vapor sensor development through their Small Business Innovation Research[6] programs. Meanwhile, the IAGOS thin film sensor has performed well across its fleet of specially equipped aircraft, but the sensors have been shown to require regular maintenance and calibration that would be infeasible for routine global airline operations at scale.

> Finding: Current in situ sensor systems lack the reliability and calibration stability necessary for widespread deployment across the commercial fleet. Accuracy of 1–2 ppm is desired down to a lower detection limit of 20–30 ppm. Sensors need to integrate with existing aircraft data downlink systems and be certified for new and existing aircraft.

Analysis of data from the IAGOS and WVSS networks provides important insights into the utility of high-accuracy water vapor measurements from commercial aircraft, particularly when assessed in a statistical and climatic sense. However, these networks cover only a small fraction of the global flight routes with a concentration of observations over North America and Europe (Figure 3-2). There is a need to extend these observations to include observations in oceanic flight corridors as well as the rest of the world in order to improve global model predictions of ISSRs and contrail formation and persistence.

---

[5] Advanced Research Projects Agency–Energy, "Document - Pre Trails Project Descriptions," https://arpa-e.energy.gov/document/pre-trails-project-descriptions, accessed December 1, 2024.

[6] NASA, "Humidity Probe for Contrail-Cirrus Avoidance," TechPort, https://techport.nasa.gov/projects/154493, accessed December 1, 2024.

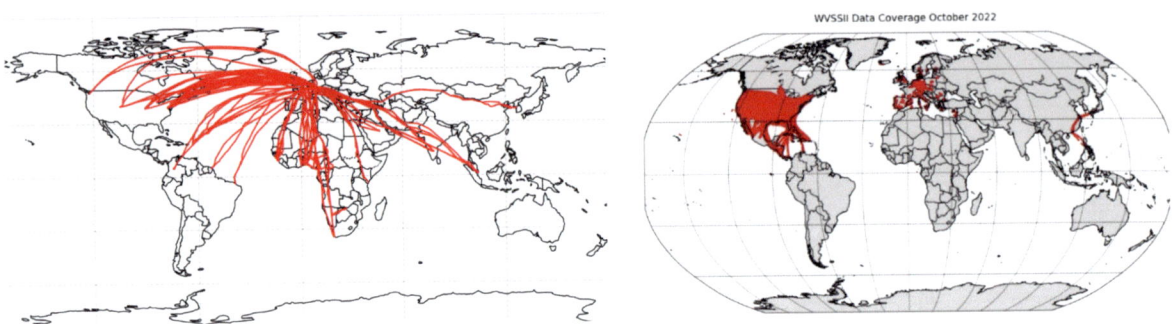

FIGURE 3-2 Map of In-Service Aircraft for a Global Observing System (*left*) and Water Vapor Sensing System (*right*) flight routes.
SOURCE: Courtesy of Canadian Meteorological Center, Environment and Climate Change Canada.

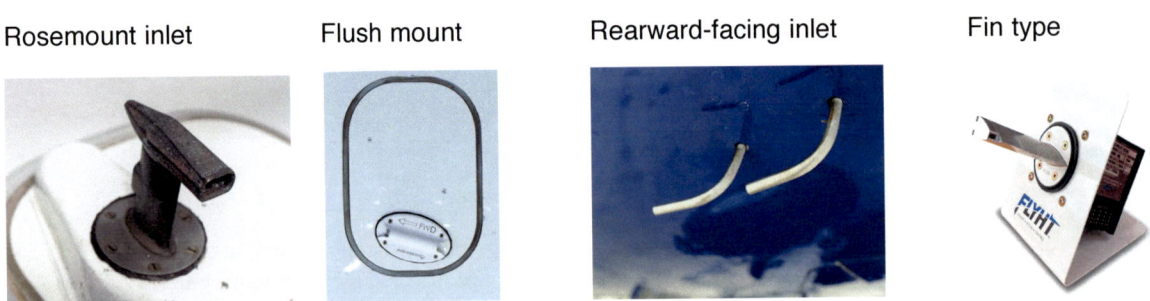

FIGURE 3-3 Example water vapor inlets that can be mounted to commercial aircraft.
SOURCES: Courtesy of FAAM Airborne Laboratory/National Centre for Atmospheric Science and FLYHT Aerospace Solution, https://flyht.com/weather-sensors/tamdar.

Finding: Deployment of a large number of science-quality water vapor sensors across the commercial fleet would be valuable to model cruise-level ice supersaturation for NOAA, FAA, and other national and international stakeholders.

Airframe integration is also another driving consideration, because a suitable water vapor sensor would need to be positioned forward of any doors or other sources of contamination (e.g., in the forward avionics bay) and avoid the need for pumps or other mechanical systems that would wear out quickly and/or increase the maintenance and complexity of the instrument. Flush-mounted, aspirating inlets like that used for the WVSS-II are a promising approach, but also may be susceptible to ingesting water or other contaminants (Figure 3-3).

These considerations were highlighted in a recent WMO workshop on aircraft-based water vapor measurements that took place in December 2023[7] and the International Air Transport Association "In-Depth" report *Aviation Contrails and Their Climate Effect Tackling Uncertainties and Enabling Solutions* (IATA 2024).

The information from the temperature, pressure, and water vapor sensors allows for theoretical calculation of whether a contrail will form and persist (to within the sensor uncertainties), but there are other engine and fuel specific parameters that also feed into the calculation of the Schmidt–Appleman criterion that may not be known. Thus, there is also some value to be gained from contrail imaging cameras (e.g., visible and/or infrared) to validate

---

[7] World Meteorological Organization, "Workshop on Aircraft-Based Water Vapour Measurement for Aviation Application," December 7–8, 2023, https://community.wmo.int/en/meetings/wvm-workshop-2023, accessed December 1, 2024.

ATMOSPHERIC MEASUREMENTS

if the aircraft is actually forming a contrail in order to account for these additional sources of uncertainty. However, cameras are limited in their ability to determine contrail persistence, and there would be a need for onboard imaging processing and contrail detection in order to limit the necessary bandwidth of information to be downlinked (Figure 3-4). Additionally, there are workload concerns associated with involving the flight crew in monitoring, reporting, or validating contrail formation via in-flight sensors or camera observations, and it is paramount that any crew involvement not impede safety priorities.

Widespread deployment of water vapor sensors, and also possibly cameras, across the commercial fleet serves two important objectives: first, the data can be assimilated into numerical weather prediction models to constrain their prediction skill for forecasting ISSRs, and second, the data can be used to diagnose contrail formation for individual aircraft flights. The key is to achieve an acceptable global coverage while also minimizing the number of sensors that need to be installed on aircraft. Observations are particularly needed for remote oceanic regions where current information is most scarce. Of course, the key question then is how many aircraft will need to incorporate new water vapor sensors in order to meaningfully inform these contrail management objectives. NASA is well positioned to employ its global atmospheric models (discussed in Chapters 4 and 5) to answer this question

FIGURE 3-4 Example of tail-mounted visible camera imagery displayed on a seat-back screen showing a contrail forming behind the engines of an Airbus A350-900 with Rolls-Royce Trent XWB engines flying westbound over the North Atlantic Ocean on September 14, 2024. An operational system would likely not transmit camera data, but only an indication that the aircraft is generating a contrail.

through observing system simulation experiments that study the best way to maximize coverage with the minimum capital and operational expenditure.

> Finding: The number and distribution of necessary sensors deployed across the fleet have yet to be optimized and will depend on the operational avoidance and verification goals and whether these are to be realized at the individual flight or fleet level.

> **Recommendation: NASA should support observing system simulation experiments to define widespread water vapor sensor deployment to best inform contrail forecasts systems and individual verification and avoidance efforts.**

In Situ and Remote Sensing Measurements on Research Aircraft

Recent NASA Science Mission Directorate (SMD) and Aeronautics Research Mission Directorate flight campaigns to study aircraft engine emissions and contrail formation at cruise altitudes (e.g., 2024 NASA CODEX, 2023 NASA Boeing ecoDemonstrator, 2018 NASA DLR ECLIF2/ND-MAX) using NASA's flying laboratories serve as ideal test beds, both for obtaining high-quality atmospheric data to validate models, as well as for flying new and in-development water vapor sensors to assess their performance under real-world conditions. The NASA Airborne Science Program's state-of-the-art water vapor measurement capabilities in the Science Directorate at Langley Research Center include the Diode Laser Hygrometer[8] (DLH) and the High Altitude Lidar Observatory[9] (HALO). Both offer advantages in terms of accuracy and precision over the small sensors discussed earlier due to the open-path nature of their measurement techniques that do not require extractive inlets or tubing. This is because water tends to stick to surfaces, so care must be taken with inlets and internal tubing (e.g., heating and flushing with dry gases) to obtain comparable levels of accuracy with cabin-mounted instruments. However, the open-path research instruments also have external lasers that may pose eye safety concerns or bulky standoffs that would be impractical for integration on commercial aircraft. An example of such standoffs is shown in Figure 3-5 for the short-path DLH. However, these are not concerns for NASA research aircraft.

The traditional DLH instrument laser path extends from a window-mounted transceiver to a retroreflector mounted on a wingtip or engine nacelle (Diskin et al. 2002), and a new short-path version of the instrument has been recently developed and flown that transmits between two arms on a single window blank. As such, the instrument makes a local, in situ measurement of water vapor in close proximity to the aircraft. The nadir-pointing HALO instrument laser extends from the aircraft altitude to the surface and remotely senses a vertically resolved water vapor curtain below the aircraft from light backscattered toward the aircraft (Carroll et al. 2022). Both instruments use differential absorption techniques at multiple absorption spectral lines (near 935 and 1404 nm for HALO and DLH, respectively). The vertical resolution of the HALO lidar water vapor curtain (e.g., Figure 3-6) is particularly advantageous for characterizing the vertical extent of ISSRs typically using model reanalysis temperature profiles that can be further constrained by nearby dropsonde or radiosonde observations.

NASA's fleet of research aircraft available through the SMD Earth Science Division Airborne Science Program[10] have multiple aircraft that would be suitable both for evaluating new airborne temperature and humidity sensors alongside either the in situ DLH instrument or while flying in tandem with the HALO remote sensor. The DC-8 flying laboratory was retired in 2024, and NASA is currently standing up a Boeing 777-200ER to be its replacement (expected start of service in 2026). Multiple Gulfstream aircraft (currently G-III and G-V variants) are also available that have the requisite ceiling, payload, and range capabilities for studying contrails, upper tropospheric aerosols, and ice supersaturation, likely at substantially lower operating costs than the relatively larger 777-200ER.

---

[8] NASA, "Diode Laser Hygrometer," Airborne Science Program, https://airbornescience.nasa.gov/instrument/DLH, accessed December 1, 2024.

[9] NASA, "High Altitude Lidar Observatory," Airborne Science Program, https://airbornescience.nasa.gov/instrument/HALO, accessed December 1, 2024.

[10] NASA, "Airborne Science Program," https://airbornescience.nasa.gov, accessed December 1, 2024.

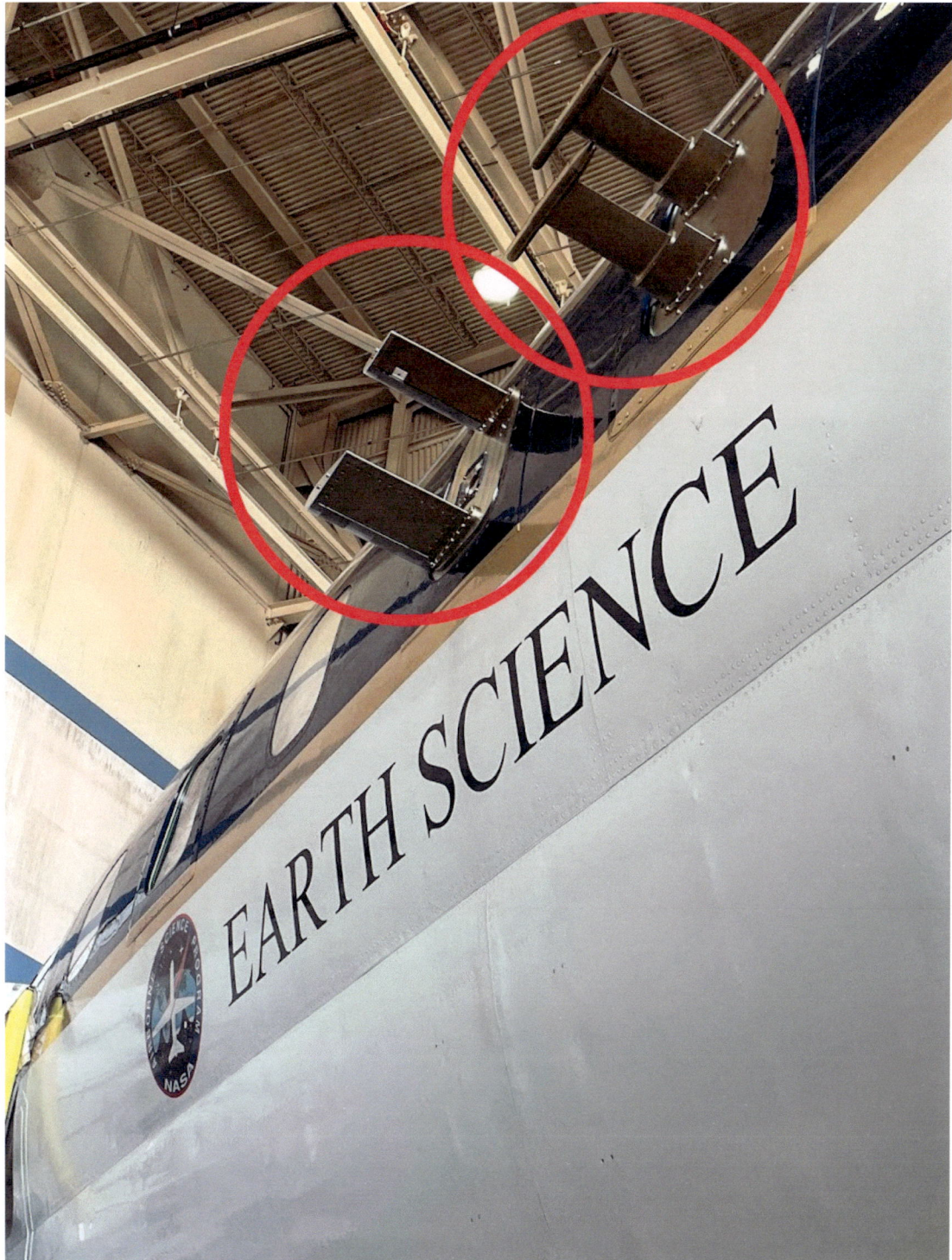

FIGURE 3-5 Short- and long-path Diode Laser Hygrometer transceivers mounted on a window plate (circled bottom left) just forward of the air sampling inlets for in-cabin WVSS-II, and Aerodyne water vapor sensors (circled top right) as configured for the 2023 NASA Boeing ecoDemonstrator Emissions Flight Test on the recently retired NASA DC-8 flying laboratory.
SOURCE: Courtesy of NASA/Richard Moore.

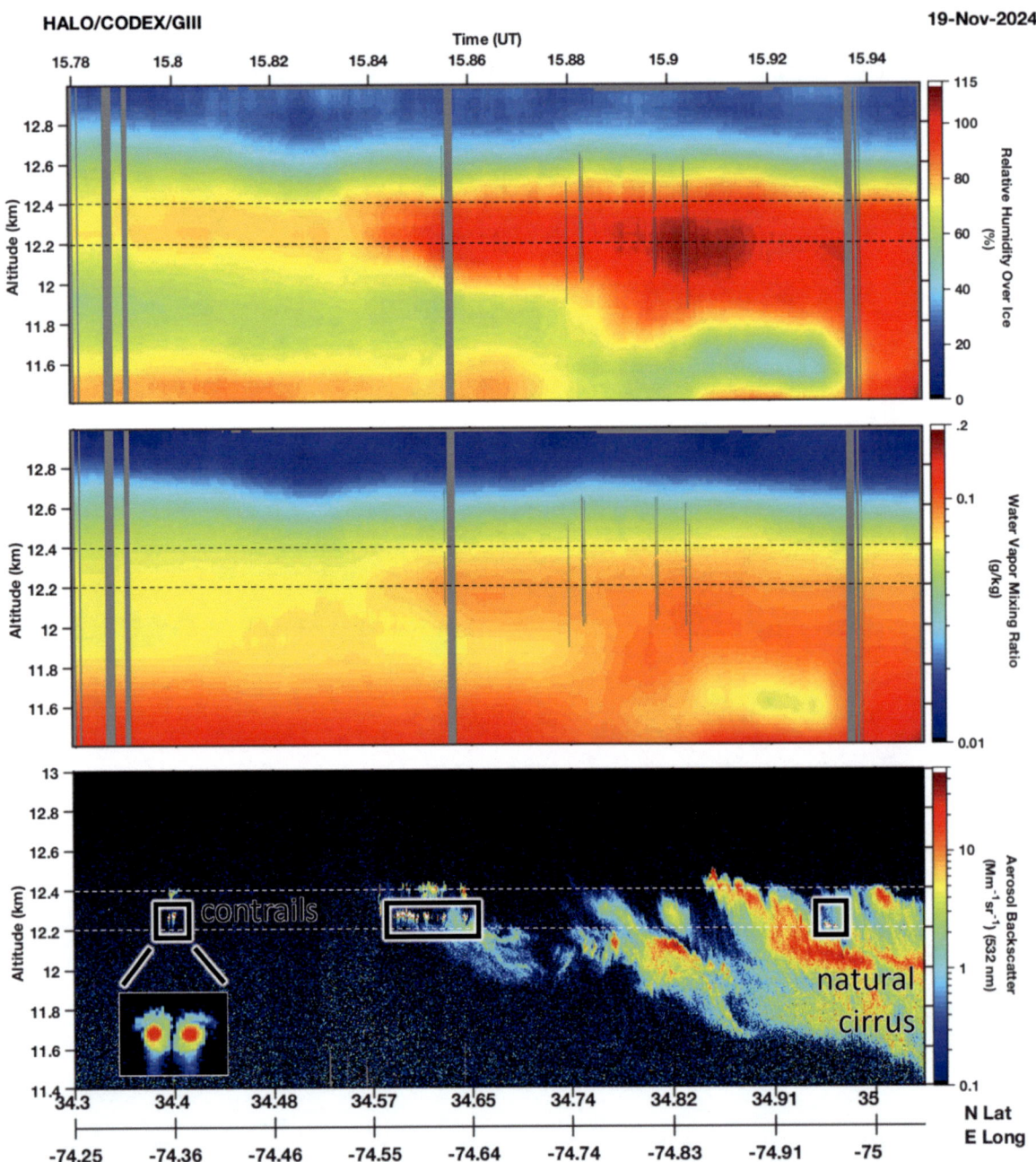

FIGURE 3-6 An example of calculated relative humidity with respect to ice curtain (top), measured water vapor mixing ratio curtain (middle), and measured aerosol backscatter coefficient curtain (bottom) obtained from the NASA High-Altitude Lidar Observatory (HALO). Relative humidity was computed using a dropsonde temperature profile launched outside of the cirrus region. Black boxes and inset denote the cross-section images of contrail backscatter coefficients measured from behind and above the GE Aerospace Flying Test Bed.
SOURCE: Courtesy of NASA/Amin Nehrir.

In addition to NASA, there are a number of other government and industry research and engineering test bed aircraft that could be suitable for evaluating water vapor sensor performance in flight. Examples of some government-funded aircraft include the DLR HALO G-V and Falcon research aircraft, the National Center for Atmospheric Research G-V, and the United Kingdom–based FAAM BAe-146. Examples of some industry-funded aircraft include the Boeing ecoDemonstrators, the GE Aerospace Flying Test Bed, the Pratt & Whitney Flying Test Bed, and Airbus and airline demonstrators. While acknowledging that flight testing is the gold standard for evaluating sensor performance under real-world conditions, it is also worthwhile and less expensive to evaluate sensors and intercompare their performance in the laboratory. Numerous government, academic, and industry research institutions, including NASA, have the necessary laboratory equipment and expertise to carry out such intercomparisons, which may be a valuable prerequisite to flight testing.

In November 2024, NASA and GE Aerospace conducted a joint flight test series called the Contrail Optical Depth Experiment (CODEX) to demonstrate how the NASA HALO lidar could be used to evaluate promising sensors and test forecast tools to identify ISSRs and contrail formation regions. The HALO lidar was integrated on the Langley Research Center G-III and measured vertical profiles of atmospheric water vapor mixing ratio (shown in Figure 3-7) and aerosol extinction and backscatter coefficients below the aircraft. The G-III chased the GE Aerospace 747 Flying Test Bed, which was equipped with an in situ WVSS-II water vapor instrument, allowing comparisons to be made between the two different sensors as well as dropsondes launched from the G-III to further constrain the atmospheric temperature and water vapor profiles. The vertical information provided by the lidar allowed for real-time optimization of the in situ aircraft flight altitude to position the aircraft within the ISSR. The CODEX project highlights one promising, two-aircraft concept of operations to study the distribution of ISSRs and evaluate forecast model predictions as well as support the testing of new water vapor sensors being developed for deployment across the commercial aircraft fleet (Figure 3-8).

FIGURE 3-7 Artist rendering of the concept of operations for the 2024 NASA–GE Aerospace Contrail Optical Depth Experiment (CODEX).
SOURCE: Courtesy of NASA/Tim Marvel.

FIGURE 3-8 In November 2024, NASA conducted a contrails flight campaign in collaboration with GE Aerospace and its contrail prediction subsidiary, SATAVIA. The campaign involved flying a NASA Langley Research Center Gulfstream III aircraft equipped with instruments to measure contrails produced by the Boeing 747 Flying Test Bed normally used as part of GE Aerospace's propulsion test programs. Contrails research requires analyzing both older, legacy aircraft (i.e., more than 10–20 years old) as well as newer aircraft. It also requires assessing the impacts of traditional aviation fuels as well as synthetic aviation fuels now in development.
SOURCES: (a) Courtesy of NASA/David Bowman; (b–d) Courtesy of NASA/Richard Moore.

Finding: NASA is uniquely positioned to test novel in situ temperature and humidity sensors using existing Science Mission Directorate Airborne Science Program aircraft and state-of-the-art research instruments.

Recommendation: NASA should support the development, testing, and certification of advanced and accurate commercial-aircraft-capable humidity and temperature sensors for contrail-forming regions as well as onboard contrail-detecting cameras and automated contrail-detection image-recognition algorithms.

Satellite and Ground-Based Remote Sensing Measurements

Spaceborne remote sensing instruments provide important additional information of atmospheric temperature and humidity profiles and contrail formation/evolution across regional-to-global scales. Skyward-pointing cameras on the ground bridge these scales to fill in the local-to-regional context and allow for validation of contrail formation and persistence.

Satellite sounding data have, to date, not been widely explored for contrail applications but hold great promise for improving numerical weather predictions of ISSRs. Currently, these instruments are limited to polar-orbiting satellites in low Earth orbit (LEO), but the next generation of geostationary (GEO) satellites set to launch in this decade by Japan and Europe (Japan Meteorological Agency and European Organisation for the Exploitation of Meteorological Satellites, respectively), and in the next decade by the United States (National Oceanic and Atmospheric Administration) promise to combine the high vertical resolution of today's LEO satellites with the high temporal and horizontal resolution achievable from GEO. China already has a GEO sounder in space, but the data and ability to collaborate is complicated by the geopolitical landscape. The expected geographical coverage and nominal launch years are shown in Figure 3-9.

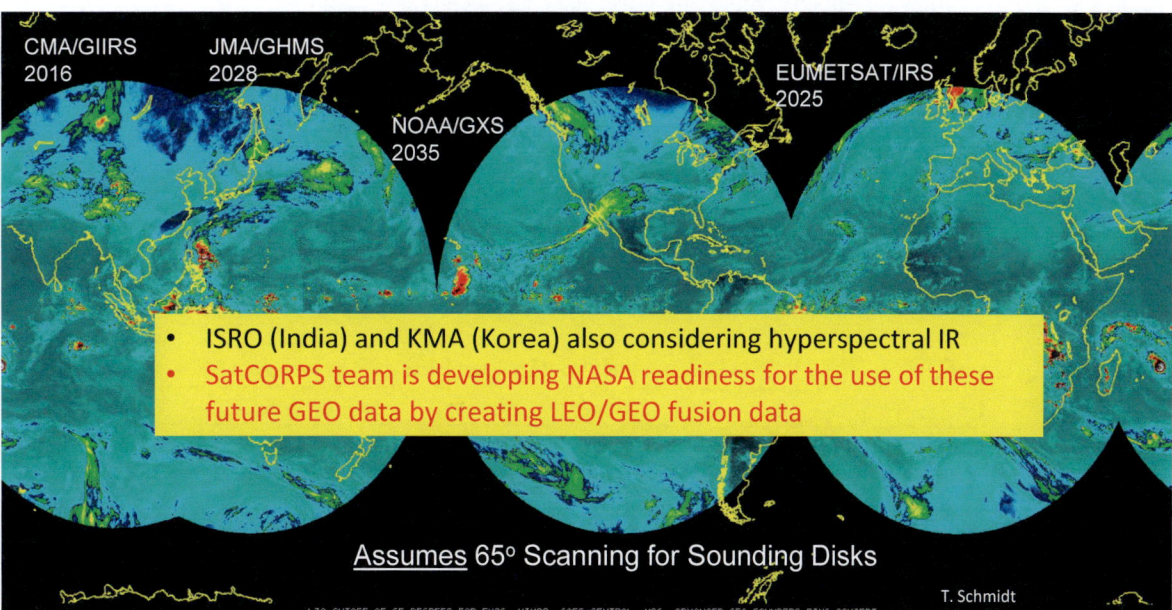

FIGURE 3-9 Current and planned geostationary satellite hyperspectral infrared sounders for remotely sensing atmospheric water vapor and temperature.
SOURCE: Courtesy of W. Smith, Jr., NASA.

Hyperspectral infrared sounders have thousands of channels that provide ~11–13 independent pieces of information on temperature and water vapor as compared to 2 independent pieces from the GOES Advanced Baseline Imager. Until the geostationary constellation comes online, there is a need to explore how today's LEO and GEO data sets could be fused to provide high-resolution (2 km, 30 minute) temperature and humidity vertical profiles from satellites. (See Table 3-2.) Essentially, this data fusion technique extrapolates the high-vertical-resolution data from LEO sounders to the high horizontal and temporal resolution of the GEO imagers. Despite these potential advances, it is clear from Figure 3-9 that major observational gaps will remain, including critical Atlantic and Pacific Ocean flight corridors. These blind spots—along with the improved, but still relatively coarse, vertical resolution of water vapor observations from space—underscore the importance of also advancing in situ sensors on the commercial aircraft fleet. The two systems could nicely complement each other by using the satellite temperature and humidity data to benchmark the performance of individual in situ water vapor sensors on the commercial fleet and identify specific sensors for closer examination and possible maintenance if they exceed a difference threshold from the satellite benchmark. The application of LEO-GEO fusion and model assimilation for ISSR and contrail predictions is a unique and important research capability of NASA Langley Research Center's Satellite Cloud and Radiative Property Retrieval System within SMD that needs to be further explored.

Finding: Next-generation geostationary satellite sounders that may be relevant for flight-level temperature and humidity as well as tracking persistent contrails will launch over the coming decade.

Finding: NASA fills an important research role within the United States in developing and demonstrating satellite data products and assimilating these products into models that can eventually be deployed to operational agencies (e.g., NOAA, FAA).

MEASUREMENTS OF CONTRAILS AND CONTRAIL CIRRUS

Both LEO and GEO satellite visible and infrared imagers have been used extensively to detect linear contrail features with early retrieval methods employing edge detection techniques based on 11–12 micron (or similar wavelength) infrared brightness temperature difference thresholds (e.g., Duda et al. 2013, 2019; Mannstein et al. 1999). Wang et al. (2024) incorporated multiple wavelengths and were able to detect contrails with satellite image masks designed for identifying dust (e.g., for the scene over northern France shown in Figure 3-10). More recent studies are employing artificial intelligence and machine learning to detect persistent contrails (e.g., Meijer et al. 2022; Ng et al. 2024; Zhang et al. 2018). Automated tracking algorithms have also been developed to monitor the evolution of the linear contrail features as they spread, drift, and grow into diffuse contrail cirrus (Vázquez-Navarro et al. 2010, 2015); however, the contrail cirrus become increasingly difficult to distinguish from natural cirrus after only a few hours. Consequently, most past research has focused on the early persistent contrail stages and

TABLE 3-2 Current Operational Satellite Instruments Available for LEO-GEO Data Fusion

Function	Orbit	Pro/Con	Sensor	Satellite Series	Domain
Hyperspectral IR Sounders	Polar	High vertical resolution (~1–2 km), but with 14 km horizontal and ~6 hr temporal resolution	CrIS	NOAA JPSS	Global
			IASI	EUMETSAT MetOp	Global
Microwave Sounders	Polar	Penetrates clouds	ATMS	NOAA JPSS	Global
			AMSU-A	EUMETSAT MetOp	Global
			MHS	EUMETSAT MetOp	Global
VIS/IR/NIR Imagers (global coverage)	GEO	High horizontal (~2 km) and temporal (~5 15 min) resolution, but with 5–10 km vertical resolution	ABI	NOAA GOES	Americas
			AHI	JMA Himawari	Asia/Pacific
			SEVIRI	EUMETSAT MSG	Europe

SOURCE: Courtesy of W. Smith, Jr., NASA.

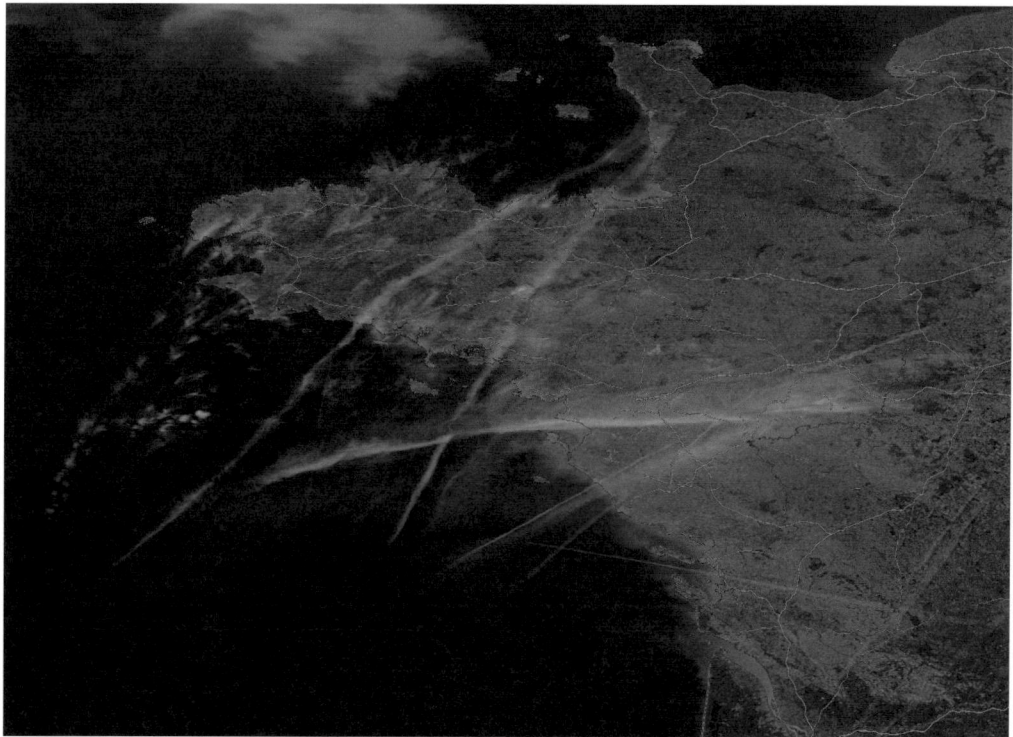

FIGURE 3-10 MODIS (Moderate Resolution Imaging Spectroradiometer) Terra Imagery of contrail cirrus forming over northwestern France on June 23, 2020.
SOURCE: Image from NASA Worldview, https://worldview.earthdata.nasa.gov/?v=-15.732016291519663,40.44025621196383, 10.529397084727265,53.5504461708871&l=Reference_Labels_15m(hidden),Reference_Features_15m,Coastlines_15m,VIIRS_NOAA21_CorrectedReflectance_TrueColor(hidden),VIIRS_NOAA20_CorrectedReflectance_TrueColor(hidden),VIIRS_SNPP_CorrectedReflectance_TrueColor(hidden),MODIS_Aqua_CorrectedReflectance_TrueColor(hidden),MODIS_Terra_CorrectedReflectance_TrueColor&lg=true&t=2020-06-23-T17%3A34%3A16Z.

much less satellite research has been done on contrail cirrus despite its importance to climate. Examination of the contrail scene in Figure 3-10 with obvious persistent contrails begs the question of what the counterfactual scene would look like if these aircraft had not flown those routes or altered their trajectories to avoid forming contrails (the subject of Chapter 6 of this report)—Would natural cirrus have formed anyway? Alternatively, would these mid-morning contrails have been avoided, but the undepleted regional water vapor budget cause more intense contrails later in the day? Absent unexpected, large-scale disruptions in air traffic, such as the September 11, 2001, terrorist attacks in the United States or the COVID-19 pandemic, such questions are currently only within the purview of modeling studies. Analyses of these events also yield mixed results, with some studies attributing changes in daily temperature range to the reduction in contrails (Travis et al. 2002, 2004), while other studies highlight confounding variability in natural cloud cover, temperature, humidity, and winds as alternate explanations that cast doubt on attempts to draw causal inferences to contrail effects (Hong et al. 2008; Kalkstein and Balling 2004). As the research community is considering operational rerouting or avoidance trials, there is an emerging need for suitable observational validation tools to quantify the (hopefully beneficial) perturbation associated with such a flight trial. Such flight trial validations need to be robust to changes in the natural atmospheric background state before, during, and after the study period that might bias a difference-based calculation of the contrail impact.

In addition, image-based identification of contrail formation and persistence (or absence), combined with aircraft position information, provides a coarse, indirect validation of model predictions of upper tropospheric temperature and humidity and whether the Schmidt–Appleman criterion and ice-supersaturation conditions are satisfied. Such techniques are also applicable to ground-based observers. Automated camera systems integrated with ADS-B are a promising method for obtaining high-quality contrail observational data sets that could be used to train and improve models and satellite retrieval algorithms, but these are highly localized and are likely to lose sight of the persistent contrails as they advect out of the field of view (Low et al. 2024; Schumann et al. 2013). Tapping into existing networks such as the Global Meteor Network[11] would greatly expand the statistical power of the contrail data set while also reducing barriers to entry in operating new cameras in target flight corridors.[12]

Ground-based observations can also be made from mobile phone cameras with integrated aircraft information from ADS-B, such as the augmented reality mode of the popular FlightRadar24 app (Figure 3-11). Such activities lend themselves to citizen science investigations and educating the public about aviation climate impacts (Colón Robles et al. 2020). A key challenge to these efforts is having access to a reliable database of aircraft position information and to create a curated, publicly accessible database for these contrail observations. NASA is well positioned to overcome these obstacles given its extensive experience with Earth Science data systems and integration into the Global Learning and Observations to Benefit the Environment[13] (GLOBE) Cloud Observations Program.

Finding: Satellite and ground-based imagers and lidars, automated detection algorithms, and flight trajectory information from ADS-B together are important tools for validating contrail model predictions and contrail-free zones.

Finding: Artificial intelligence and machine learning are becoming more widely used for automated contrail observation and detection from satellite and ground-based imagery.

Recommendation: NASA should support satellite remote sensing research for diagnosing persistent contrails and ice-supersaturated regions in order to develop readiness for the next-generation geostationary sounders and imagers.

MEASUREMENTS OF ATMOSPHERIC PARTICLES

Aerosol particles serve as the cloud condensation nuclei on which contrails form as the water vapor–rich aircraft engine exhaust plume cools. These particles may be emitted from the aircraft and/or engine exhaust or may already exist in the upper troposphere from other natural and anthropogenic emissions sources (Figure 3-12). As discussed in Chapter 2, most current engine and fuel combinations used across the aviation fleet emit substantial numbers of particles that vastly outnumber the relatively few pre-existing background particles in the upper troposphere. For these "soot-rich" engines, the number of contrail ice crystals scales proportionally to the number of emitted soot particles and the ambient temperature difference relative to the Schmidt–Appleman contrail formation temperature (Kärcher 2018; Kärcher and Yu 2009). However, with the future adoption of lower-sooting engine technologies and sustainable aviation fuels containing zero or reduced fuel sulfur and aromatics in the coming decades, the role of pre-existing particles in the background atmosphere may become more important for contrail formation. Consequently, there is an emerging need to understand and characterize these background particles within and across flight corridors and to determine the extent to which these particle populations derive from terrestrial or aviation sources.

In addition, and as discussed in more detail in Chapters 5 and 6, modeling efforts to understand the climate impact of contrail cirrus must be able to reliably simulate the contrail formation and evolution in the atmosphere as well as the counterfactual atmospheric state where the aircraft did not fly. Essentially, the question being asked is whether natural cirrus clouds would have formed in the absence of an aviation-related perturbation via homogeneous or heterogeneous freezing pathways on existing aerosols. Thus, to answer this question and improve

---

[11] See the Global Meteor Network website at https://globalmeteornetwork.org, accessed December 1, 2024.
[12] Some examples of collocated ground video cameras are https://www.youtube.com/watch?v=DUIoQJUn1YU during nighttime and https://www.youtube.com/watch?v=6JxqNqtUJQc during daytime.
[13] See the GLOBE website at https://www.globe.gov, accessed December 1, 2024.

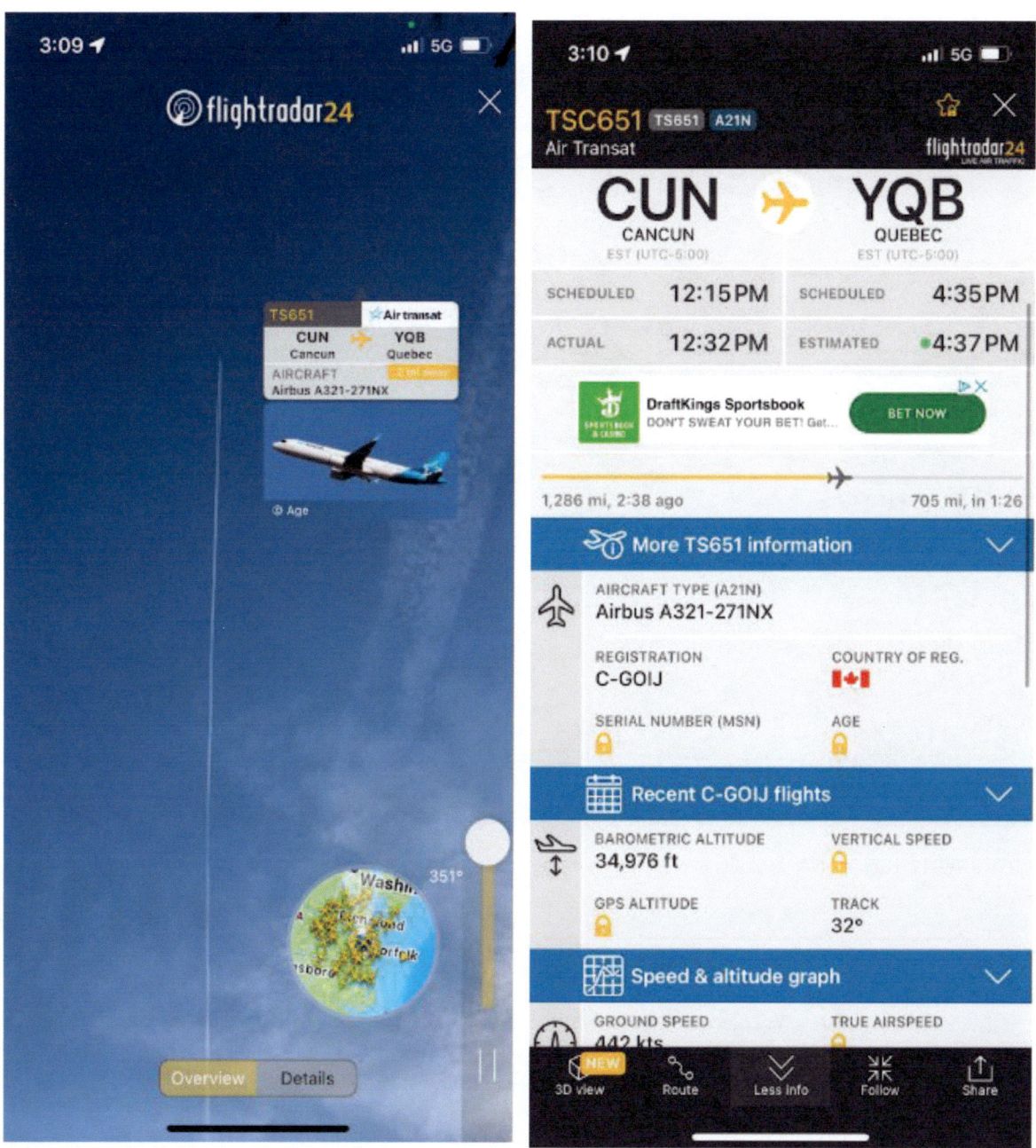

FIGURE 3-11 Ground-based sky camera image of a persistent contrail with augmented reality identification of the aircraft and ADS-B altitude that could be used to validate model predictions.
SOURCE: Courtesy of Flightradar24.com, https://www.flightradar24.com.

FIGURE 3-12 High-altitude contrails over Colorado in late 2024.

models, information is needed about the pre-existing upper tropospheric aerosol population and the presence of ice-nucleating particles.

NASA's Science Mission Directorate has conducted airborne science field campaigns to study the remote upper troposphere (e.g., the Atmospheric Tomography Mission), and many projects involve transit flights at cruise altitudes to reposition the aircraft from its base of operations to the project deployment site. Such flights are not confined to typical airline flight corridors (and may even seek to avoid congested regions) but could yield valuable information about upper tropospheric composition if they included a small, targeted payload of particle instrumentation. Current airborne aerosol research instrumentation is mature but of sufficient complexity (e.g., use of flammable working fluids and isokinetic sampling inlets) to be impractical for widespread deployment on commercial aircraft. Condensation particle counters, optical particle sizers, and aerosol mobility sizers are routinely deployed on research aircraft operated by NASA and other national and university research institutions around the world and would be suitable for characterizing the upper tropospheric aerosol. Other aerosol properties of interest (particularly in the aviation-relevant, 5–100 nm diameter size range), include the following:

- Volatile and non-volatile particle number size distributions and concentrations,
- Particle chemical composition and mixing state, and
- Cloud and ice nucleating properties.

New aerosol instrument development efforts would be valuable that decrease the size, weight, and power requirements to enable deployment of these sensors across flight campaigns that are often payload limited. In addition, current mass-based composition measurements are challenged by the small particle sizes of aircraft engine emissions (<100 nm in diameter), and instrument development efforts in this area are needed to advance the state of the art.

Finding: Upper tropospheric aerosol number concentrations and properties are poorly understood for both continental and, especially, remote marine regions.

Finding: Natural cirrus form on ice-nucleating particles and the role of aircraft engine particle emissions in altering these aerosol-cloud interactions near flight corridors is highly uncertain.

Recommendation: NASA should identify and enable a minimum set of key aerosol instruments that can be flown on multiple missions with the goal of characterizing the aerosol composition of the upper troposphere and uncovering the contribution of aviation emissions relative to other sources.

Chapter 4 discusses modeling systems that will benefit from the improved atmospheric measurement data recommended in this chapter.

REFERENCES

Afchine, A., C. Rolf, A. Costa, N. Spelten, M. Riese, B. Buchholz, V. Ebert, et al. 2018. "Ice Particle Sampling from Aircraft—Influence of the Probing Position on the Ice Water Content." *Atmospheric Measurement Techniques* 11(7):4015–4031. https://doi.org/10.5194/amt-11-4015-2018.

Appleman, H. 1953. "The Formation of Exhaust Condensation Trails by Jet Aircraft." *Bulletin of the American Meteorological Society* 34(1):14–20. https://doi.org/10.1175/1520-0477-34.1.14.

Buchholz, B., A. Afchine, A. Klein, C. Schiller, M. Krämer, and V. Ebert. 2017. "HAI, A New Airborne, Absolute, Twin Dual-Channel, Multi-Phase TDLAS-Hygrometer: Background, Design, Setup, and First Flight Data." *Atmospheric Measurement Techniques* 11:35–57. https://doi.org/10.5194/amt-10-35-2017.

Buck Research Instruments. 2024. "1011C Aircraft Hygrometer." https://www.hygrometers.com/products/1011c. Accessed December 1, 2024.

Carroll, B.J., A.R. Nehrir, S.A. Kooi, J.E. Collins, R.A. Barton-Grimley, A. Notari, D.B. Harper, and J. Lee. 2022. "Differential Absorption Lidar Measurements of Water Vapor by the High Altitude Lidar Observatory (HALO): Retrieval Framework and First Results." *Atmospheric Measurement Techniques* 15(3):605–626. https://doi.org/10.5194/amt-15-605-2022.

Colón Robles, M., H.M. Amos, J.B. Dodson, J. Bouwman, T. Rogerson, A. Bombosch, L. Farmer, A. Burdick, J. Taylor, and L.H. Chambers. 2020. "Clouds Around the World: How a Simple Citizen Science Data Challenge Became a Worldwide Success." *Bulletin of the American Meteorological Society* 101(7):E1201–E1213. https://doi.org/10.1175/BAMS-D-19-0295.1.

Diskin, G.S., J.R. Podolske, G.W. Sachse, and T.A. Slate. 2002. "Open-Path Airborne Tunable Diode Laser Hygrometer." In *Proceedings Volume 4817, Diode Lasers and Applications in Atmospheric Sensing*. International Symposium on Optical Science and Technology. https://doi.org/10.1117/12.453736.

Duda, D.P., P. Minnis, K. Khlopenkov, T.L. Chee, and R. Boeke. 2013. "Estimation of 2006 Northern Hemisphere Contrail Coverage Using MODIS Data." *Geophysical Research Letters* 40:612–617. https://doi.org/10.1002/grl.50097.

Duda, D.P., S.T. Bedka, P. Minnis, D. Spangenberg, K. Khlopenkov, T. Chee, and W.L. Smith. 2019. "Northern Hemisphere Contrail Properties Derived from Terra and Aqua MODIS Data for 2006 and 2012." *Atmospheric Chemistry and Physics* 19(8):5313–5330.

Ford, B. 2011. "Water Vapor Sensing System (WVSS-II), SpectraSensors." https://www.slideserve.com/messina/water-vapor-sensing-system-wvss-ii-powerpoint-ppt-presentation.

Helten, M., H.G.J. Smit, W. Sträter, D. Kley, P. Nedelec, M. Zöger, and R. Busen. 1998. "Calibration and Performance of Automatic Compact Instrumentation for the Measurement of Relative Humidity from Passenger Aircraft." *Journal of Geophysical Research: Atmospheres* 103(D19):25643–25652. https://doi.org/10.1029/98JD00536.

Hong, G., P. Yang, P. Minnis, Y.X. Hu, and G. North. 2008. "Do Contrails Significantly Reduce Daily Temperature Range?" *Geophysical Research Letters* 35:L23815. https://doi.org/10.1029/2008GL036108.

IAGOS (In-Service Aircraft for a Global System). 2024. "Humidity Sensor (ICH, Part of Package1)." https://www.iagos.org/iagos-core-instruments/h2o. Accessed December 1, 2024.

IATA (International Air Transport Association). 2024. *Aviation Contrails and Their Climate Effect: Tackling Uncertainties and Enabling Solutions.* Technical Report. https://www.iata.org/contentassets/726b8a2559ad48fe9decb6f2534549a6/aviation-contrails-climate-impact-report.pdf.

Kalkstein, A.J., and R.C. Balling. 2004. "Impact of Unusually Clear Weather on United States Daily Temperature Range Following 9/11/2001." *Climate Research* 26(1):1–4. http://www.jstor.org/stable/24868704.

Kärcher, B. 2018. "Formation and Radiative Forcing of Contrail Cirrus." *Nature Communications* 9(1):1824.

Kärcher, B., and F. Yu. 2009. "Role of Aircraft Soot Emissions in Contrail Formation." *Geophysical Research Letters* 36(1). https://doi.org/10.1029/2008GL036649.

Karion, A., C. Sweeney, S. Wolter, T. Newberger, H. Chen, A. Andrews, J. Kofler, D. Neff, and P. Tans. 2013. "Long-Term Greenhouse Gas Measurements from Aircraft." *Atmospheric Measurement Techniques* 6(3):511–526. https://doi.org/10.5194/amt-6-511-2013.

Kaufmann, S., C. Voigt, T. Jurkat, T. Thornberry, D.W. Fahey, R.-S. Gao, R. Schlage, D. Schäuble, and M. Zöger. 2016. "The Airborne Mass Spectrometer AIMS—Part 1: AIMS-H_2O for UTLS Water Vapor Measurements." *Atmospheric Measurement Techniques* 9:939–953. https://doi.org/10.5194/amt-9-939-2016.

Kaufmann, S., C. Voigt, R. Heller, T. Jurkat-Witschas, M. Krämer, C. Rolf, M. Zöger, et al. 2018. "Intercomparison of Mid-Latitude Tropospheric and Lower Stratospheric Water Vapor Measurements and Comparison to ECMWF Humidity Data." *Atmospheric Chemistry and Physics* 18(22):16729–16745. https://doi.org/10.5194/acp-18-16729-2018.

Low, J., R. Teoh, J. Ponsonby, E. Gryspeerdt, M. Shapiro, and M.E.J. Stettler. 2025. "Ground-Based Contrail Observations: Comparisons with Reanalysis Weather Data and Contrail Model Simulations." *Atmospheric Measurement Techniques* 18(1):37–56.

Mannstein, H., R. Meyer, and P. Wendling. 1999. "Operational Detection of Contrails from NOAA-AVHRR-Data." *International Journal of Remote Sensing* 20(8):1641–1660. https://doi.org/10.1080/014311699212650.

Marshall, C. 2024. "NOAA/NWS Aircraft-Based Observations Program." Presentation to the committee. September 12. National Academies of Sciences, Engineering, and Medicine. https://www.nationalacademies.org/event/43604_09-2024_research-agenda-for-reducing-the-climate-impact-of-aviation-induced-cloudiness-and-persistent-contrails-from-commercial-aviation-meeting-5.

Meijer, V.R., L. Kulik, S.D. Eastham, F. Allroggen, R.L. Speth, S. Karaman, and S.R.H. Barrett. 2022. "Contrail Coverage Over the United States Before and During the COVID-19 Pandemic." *Environmental Research Letters* 17(3):034039. https://doi.org/10.1088/1748-9326/ac26f0.

Meyer, J., C. Rolf, C. Schiller, S. Rohs, N. Spelten, A. Afchine, M. Zöger, et al. 2015. "Two Decades of Water Vapor Measurements with the FISH Fluorescence Hygrometer: A Review." *Atmospheric Chemistry and Physics* 15:8521–8538. https://doi.org/10.5194/acp-15-8521-2015.

Ng, J.Y.H., K. McCloskey, J. Cui, V.R. Meijer, E. Brand, A. Sarna, N. Goyal, C. Van Arsdale, and S. Geraedts. 2024. "Contrail Detection on GOES-16 ABI with the Opencontrails Dataset." *IEEE Transactions on Geoscience and Remote Sensing* 62:1–14.

Picarro. n.d. "CO_2, CH_4 and H_2O Gas Concentration Analyzer." G2301 Datasheet. https://www.picarro.com/sites/default/files/product_documents/Picarro_G2301%20Datasheet_230306.pdf. Accessed December 1, 2024.

PST (Process Sensing Technologies). n.d. "Chilled Mirror Hygrometer—Michell S8000 Remote." https://www.processsensing.com/en-us/products/s8000-remote-chilled-mirror-hygrometer.htm. Accessed December 1, 2024.

Sausen, R., S. Hofer, K. Gierens, L. Bugliaro, R. Ehrmanntraut, I. Sitova, K. Walczak, A. Burridge-Diesing, M. Bowman, and N. Miller. 2024. "Can We Successfully Avoid Persistent Contrails by Small Altitude Adjustments of Flights in the Real World?" *Meteorologische Zeitschrift* 33(1):83–98. https://doi.org/10.1127/metz/2023/1157.

Schiller, C., J.-U. Grooß, P. Konopka, F. Plöger, F.H. Silva Dos Santos, and N. Spelten. 2009. "Hydration and Dehydration at the Tropical Tropopause." *Atmospheric Chemistry and Physics* 9(24):9647–9660. https://doi.org/10.5194/acp-9-9647-2009.

Schmidt, E. 1963. *Einführung in die Technische Thermodynamik und in die grundlagen der Chemischen Thermodynamik*, 10th ed. Springer.

Schumann, U. 1996. "On Conditions for Contrail Formation from Aircraft Exhausts." *Meteorologische Zeitschrift* 5(1):4–23. https://doi.org/10.1127/metz/5/1996/4.

Schumann, U., R. Hempel, H. Flentje, M. Garhammer, K. Graf, S. Kox, H. Lösslein, and B. Mayer. 2013. "Contrail Study with Ground-Based Cameras." *Atmospheric Measurement Techniques* 6(12):3597–3612. https://doi.org/10.5194/amt-6-3597-2013.

Spichtinger, P., and M. Leschner. 2016. "Horizontal Scales of Ice-Supersaturated Regions." *Tellus B: Chemical and Physical Meteorology* 68(1):29020. https://doi.org/10.3402/tellusb.v68.29020.

Stuefer, M., and T. Gordon. 2018. *Cryogenic Frostpoint Hygrometer (CFH) Instrument Handbook*. Department of Energy Office of Science, ARM Climate Research Facility. https://www.arm.gov/publications/tech_reports/handbooks/cfh_handbook.pdf.

Thornberry, T.D., A.W. Rollins, R.S. Gao, L.A. Watts, S.J. Ciciora, R.J. McLaughlin, C. Voigt, B. Hall, and D.W. Fahey. 2013. "Measurement of Low-PPM Mixing Ratios of Water Vapor in the Upper Troposphere and Lower Stratosphere Using Chemical Ionization Mass Spectrometry." *Atmospheric Measurement Techniques* 6:1461–1475. https://doi.org/10.5194/amt-6-1461-2013.

Thornberry, T.D., A.W. Rollins, R.S. Gao, L.A. Watts, S.J. Ciciora, R.J. McLaughlin, and D.W. Fahey. 2015. "A Two-Channel, Tunable Diode Laser-Based Hygrometer for Measurement of Water Vapor and Cirrus Cloud Ice Water Content in the Upper Troposphere and Lower Stratosphere." *Atmospheric Measurement Techniques* 8:211–224. https://doi.org/10.5194/amt-8-211-2015.

Travis, D., A. Carleton, and R. Lauritsen. 2002. "Contrails Reduce Daily Temperature Range." *Nature* 418:601. https://doi.org/10.1038/418601a.

Travis, D.J., A.M. Carleton, and R.G. Lauritsen. 2004. "Regional Variations in U.S. Diurnal Temperature Range for the 11–14 September 2001 Aircraft Groundings: Evidence of Jet Contrail Influence on Climate." *Journal on Climate* 17:1123–1134. https://doi.org/10.1175/1520-0442(2004)017<1123:RVIUDT>2.0.CO;2.

Vance, A.K., S.J. Able, R.J. Cotton, and A.M. Woolley. 2015. "Performance of WVSS-II Hygrometers on the FAAM Research Aircraft." *Atmospheric Measurement Techniques* 8(3):1617–1625. https://doi.org/10.5194/amt-8-1617-2015.

Vázquez-Navarro, M., H. Mannstein, and B. Mayer. 2010. "An Automatic Contrail Tracking Algorithm." *Atmospheric Measurement Techniques* 3(4):1089–1101. https://doi.org/10.5194/amt-3-1089-2010.

Vázquez-Navarro, M., H. Mannstein, and S. Kox. 2015. "Contrail Life Cycle and Properties from 1 Year of MSG/SEVIRI Rapid-Scan Images." *Atmospheric Chemistry and Physics* 15(15):8739–8749. https://doi.org/10.5194/acp-15-8739-2015.

Wang, X., K. Wolf, O. Boucher, and N. Bellouin. 2024. "Radiative Effect of Two Contrail Cirrus Outbreaks Over Western Europe Estimated Using Geostationary Satellite Observations and Radiative Transfer Calculations." *Geophysical Research Letters* 51(7):e2024GL108452. https://doi.org/10.1029/2024GL108452.

Zhang, G., J. Zhang, and J. Shang. 2018."Contrail Recognition with Convolutional Neural Network and Contrail Parameterizations Evaluation." *Scientific Online Letters on the Atmosphere* 14:132–137. https://doi.org/10.2151/sola.2018-023.

Zondlo, M.A., M.E. Paige, S.M. Massick, and J.A. Silver. 2010. "Vertical Cavity Laser Hygrometer for the National Science Foundation Gulfstream-V Aircraft." *Journal of Geophysical Research: Atmospheres* 115(D20). https://doi.org/10.1029/2010JD014445.

4

Contrail Modeling Systems

The goal of a contrail modeling system is to simulate the formation, persistence, and evolution of contrails and contrail-generated cirrus as well as their radiative impacts. Simulating contrails typically requires a set of multiple models for different aspects of the problem, hence a "modeling system." For an end-to-end system, one starts with a model for the engine exhaust that accounts for the combustor conditions and technology, the fuel type, and the operating conditions for each flight segment, as described in Chapter 2. Chapter 3 on measurements reflects the emphasis on needing observations to feed the models. In addition, there are only a handful of current contrail models and this chapter does not attempt to describe these models in detail as they are previously published in the literature. Chapters 5 and 6 provide more specifics about how the models will be used and what simulations are needed. This chapter describes the modeling systems that are initialized with the characterization of the exhaust at the exit plane of the engines and then simulate the initial formation of contrails behind the aircraft and their evolution and dissipation. Engine emissions are placed into an exhaust plume model that includes the airflow around the airframe and accounts for the pressure and temperature fluctuations plus vertical motion of the aircraft's wake vortices as they collect the engine exhaust and descend 100 m or more, all the while mixing the vortices with the ambient atmosphere. After the dynamical energy in the vortex dissipates, the final mixing of the exhaust plume and any contrails into the atmosphere (e.g., into general pollution layers or cirrus clouds) is controlled by the large-scale atmospheric wind shear and static stability. Aircraft engine exhaust consists of gases and particles that can continue to react with each other or the ambient atmosphere once they leave the engine exit plane. Gases can be converted to other species or condense to existing or new particles. Most importantly here, the particles can condense ambient water vapor to form initial contrail ice particles (<10 micron diameter), which can grow into ice crystals (20–100 micron diameter). Like natural cirrus clouds, these ice crystals will eventually sublimate through gravitational settling into drier regions or large-scale atmospheric subsidence and warming. Contrail modeling systems need to include the dynamical mixing of the aircraft wake with the ambient atmosphere plus the chemistry and microphysics of the particles.

WAKE VORTEX MODELS

A number of research groups have developed models of the wakes and contrails from jet aircraft.[1] In addition, there have been detailed observations of the wake vortex and contrails that can test these models (Gayet et al.

---

[1] See, for example, Fritz et al. (2020), Kärcher et al. (2015), Lewellen and Lewellen (2001), Paoli et al. (2013), Paugam et al. (2010), Picot et al. (2015), Sussmann and Gierens (1999), and Unterstrasser et al. (2008).

2012; Schumann et al. 2013). Many studies have shown that the intensity of contrails (i.e., their radiative impact on climate) depends on not simply the Schmidt–Appleman criterion (ice supersaturation ratio, designated as relative humidity over ice, RHi >100 percent; Appleman [1953]), but also other factors controlling the wake vortex such as the jet-to-wing-span ratio, aircraft weight and number of engines, ambient wind shear, and stability (Llewellen and Llewellen 2001; Paoli et al. 2013; Saulgeot et al. 2023).

Vortex dynamics follows several stages, as does the microphysics controlling the contrail ice particles. Initial visible contrails form from the expanding exhaust of each engine within 1 s. These contain all the exhaust products, especially the soot, engine oil, and trace metal particles that are seeds for the contrail ice particles, and collectively, these descend from the aircraft's downwash. After about 20 s, the pair of rotating wing-tip vortices begin to sweep up the exhaust and contrail ice, redistributing it in rotating vortices ranging from flight level to several hundred meters below (see Lewellen and Lewellen 2001). After about 400 s, the vortices decay and the remaining non-uniform distribution of ice particles enters the mature phase where the plume is dispersed by large-scale atmospheric processes (wind shear, turbulence, sublimation). Typically, the contrail ice particles gravitationally settle, falling through the atmosphere, and sublimating about 700 m below flight level (Schumann et al. 2015). The vortex phase is the most complex for particles and ice physics: the plume cools, and formation of sulfate aerosols, freezing on solid nuclei, condensation, heterogeneous nucleation, and coagulation of particles occurs. In the presence of a sufficient number of soot aerosols, heterogeneous freezing of ice occurs, but homogeneous freezing of sulfate particles might take place if soot concentrations are low enough. Organic compounds are also present in the exhaust and, like sulfate, can condense on the non-volatile soot particles or mix with sulfate particles. While these small particles can be treated as passive tracers within the bulk air flow in plume models, contrail ice has its own dynamical transport.

GLOBAL CLIMATE MODELS

The mature phase of the exhaust plume is described as the diffusion regime where the large-scale atmospheric structure defined by wind shear and static stability control the final evolution of the exhaust plume and any contrails (e.g., Fritz et al. 2020; Paoli et al. 2013), and in this regime, mixing of compounds in the plume with ambient air compounds becomes important. In addition, the plume is advected (i.e., transferred) along with the ambient air. Ice crystal growth in this regime takes place primarily from interaction of the exhaust particles and contrail ice with the ambient water vapor.

The climate impacts of contrails and contrail-induced cirrus all come from this mature phase in which the contrail vortex model is linked with a global-scale weather forecasting or climate model. These global models calculate the integrated radiative forcing and dehydration impacts of a contrail and place it in the context of the background atmosphere (water vapor, clouds, surface albedo).[2]

Global climate models (GCMs) are used directly for predicting contrail impacts (Bock and Burkhardt 2016; Burkhardt et al. 2010; Chen and Gettelman 2016; Gettelman et al. 2021), but these do not resolve clouds and individual contrails, instead generating them statistically in each of their large, 100-km grid cells and letting them evolve statistically according to the meteorological conditions and the cloud physics of the climate model. The GCM derives its initial plume parameters from observations and plume models. The advantage of a GCM is a closed mass budget wherein the evolution of water in a cloud will pull water from the ambient atmosphere and ensure proper interaction with existing and subsequent cloud formation. Climate models also have closed energy budgets and can be used to predict the radiative effects of contrails (given their parameterized cloud microphysics).

CONTRAIL PLUME MODELS

Other examples of contrail modeling involve calculation of individual flights within an atmosphere defined by current weather prediction systems. Examples are the contrail cirrus prediction model (CoCiP; Schumann 2012) as well as the plume model of Fritz et al. (2020) which focuses on the chemical and microphysical evolution of the

---

[2] See Chen and Gettelman (2016), Gettelman et al. (2021), and Schumann et al. (2015).

plume. Singh et al. (2024, see Tables 2 and 3) reviewed many of these models. These plume models are optimized to calculate contrails globally given identified aircraft flight paths and aircraft emissions plus the meteorological conditions from weather forecasting systems such as the European Centre for Medium Range Weather Forecasting. These models can include all global aircraft flights but must parameterize the dynamical processes described in the large eddy simulation models (e.g., Lewellen and Lewellen 2001; Paoli et al. 2017).

CoCiP is a plume model that takes exhaust particle number and forms contrails under the right conditions. The advection and evolution of the contrails is followed with a Lagrangian Gaussian plume model. Mixing and bulk cloud processes are treated quasi-analytically or with a numerical scheme. Contrails disappear when the bulk ice content sublimates (as a result of changes in the background atmosphere) or precipitates (when the ice particles are large enough). Water from the background ambient atmosphere condenses and is the primary mass of contrails and contrail cirrus; yet, when using a forecast model, the water vapor in the forecast model is not changed. This process is different when CoCiP is coupled with a climate model wherein the atmospheric humidity profile changes when ice crystals from the contrails fall to lower altitudes and release the water vapor (Schumann et al. 2015). Forecast/reanalysis models can provide the enabling background meteorology for these plume models (i.e., ice-supersaturation profiles, wind shear, stratification) but have not been designed to re-assimilate the redistribution of water vapor. While climate models can be, and have been, merged with plume models, the link with the forecast models is needed to predict specific time periods and to optimize flight paths for reduced contrail climate impacts. There is currently no seamless prediction system.

ICE-SUPERSATURATION FORECAST MODELS

One aspect of forecast models that needs improvement is the prediction of ice-supersaturation regions, including the amount of supersaturation, as these determine contrails formation, their persistence, and their optical depth (i.e., climate impact). Current forecast models are not optimized to accurately predict these regions and the amount of ice supersaturation.

SYSTEM FOR CONTRAIL PREDICTION

NASA has several modeling/observation systems that could be used to predict ice supersaturation and the formation of contrails using either National Oceanic and Atmospheric Administration–based models (e.g., RAPP through NASA Langley Research Center) or NASA's own assimilation system (GEOS, a weather forecast model optimized to assimilate NASA Earth system observations). Such systems could be used to predict contrails statistically or with an embedded plume model such as CoCiP. The GEOS system currently operates at higher resolution than most climate models (0.25° latitude, ~25 km).

Finding: Individual models of the atmosphere (weather forecasts) and contrails need to be linked into a seamless system so differing models and meteorological data can be tested for contrail prediction.

Recommendation: As part of a national strategy, NASA should support development and assessment of models for all scales of contrail prediction. These models range from wake vortex to global climate to contrail plume to ice-supersaturation forecast.

SUMMARY

There is currently no complete model system for predicting contrails. To optimize and test a contrail prediction system, such a modeling system needs to undergo verification as with any weather forecast system. Any contrail modeling system should be able to support stakeholder needs. The elements needed for development of an operational system to forecast and verify contrails are discussed in the Chapter 5.

REFERENCES AND FURTHER READING

Appleman, H. 1953. "The Formation of Exhaust Condensation Trails by Jet Aircraft." *Bulletin of the American Meteorological Society* 34(1):14–20.

Bock, L., and U. Burkhardt. 2016. "The Temporal Evolution of a Long-Lived Contrail Cirrus Cluster: Simulations with a Global Climate Model." *Journal of Geophysical Research: Atmospheres* 121(7):3548–3565.

Burkhardt, U., B. Kärcher, and U. Schumann. 2010. "Global Modeling of Contrail and Contrail Cirrus Climate Impact." *Bulletin of the American Meteorological Society* 91:479–483.

Chen, C.-C., and A. Gettelman. 2016. "Simulated 2050 Aviation Radiative Forcing from Contrails and Aerosols." *Atmospheric Chemistry and Physics* 16(11):7317–7333.

Fritz, T.M., S.D. Eastham, R.L. Speth, and S.R.H. Barrett. 2020. "The Role of Plume-Scale Processes in Long-Term Impacts of Aircraft Emissions." *Atmospheric Chemistry and Physics* 20(9):5697–5727.

Gayet, J.-F., V. Shcherbakov, C. Voigt, U. Schumann, D. Schäuble, P. Jessberger, A. Petzold, et al. 2012. "The Evolution of Microphysical and Optical Properties of an A380 Contrail in the Vortex Phase." *Atmospheric Chemistry and Physics* 12(14):6629–6643.

Gettelman, A., C.-C. Chen, and C.G. Bardeen. 2021. "The Climate Impact of COVID-19-Induced Contrail Changes." *Atmospheric Chemistry and Physics* 21(12):9405–9416.

Kärcher, B., T. Peter, U.M. Biermann, and U. Schumann. 1996. "The Initial Composition of Jet Condensation Trails." *Journal of the Atmospheric Sciences* 53:3066–3083.

Kärcher, B., U. Burkhardt, A. Bier, L. Bock, and I.J. Ford. 2015. "The Microphysical Pathway to Contrail Formation," *Journal of Geophysical Research: Atmospheres* 120(15):7893–7927.

Lee, D.S., G. Pitari, V. Grewe, K. Gierens, J.E. Penner, A. Petzold, M.J. Prather, et al. 2010. "Transport Impacts on Atmosphere and Climate: Aviation." *Atmospheric Environment* 44(37):4678–4734.

Lewellen, D.C., and W.S. Lewellen. 2001. "The Effects of Aircraft Wake Dynamics on Contrail Development." *Journal of the Atmospheric Sciences* 58(4):390–406.

Paoli, R., L. Nybelen, J. Picot, and D. Cariolle. 2013. "Effects of Jet/Vortex Interaction on Contrail Formation in Supersaturated Conditions." *Physics of Fluids* 25(5).

Paugam, R., R. Paoli, and D. Cariolle. 2010. "Influence of Vortex Dynamics and Atmospheric Turbulence on the Early Evolution of a Contrail." *Atmospheric Chemistry and Physics* 10(8):3933–3952.

Picot, J., R. Paoli, O. Thouron, and D. Cariolle. 2015. "Large-Eddy Simulation of Contrail Evolution in the Vortex Phase and Its Interaction with Atmospheric Turbulence." *Atmospheric Chemistry and Physics* 15(13):7369–7389.

Saulgeot, P., V. Brion, N. Bonne, E. Dormy, and L. Jacquin. 2023. "Effects of Atmospheric Stratification and Jet Position on the Properties of Early Aircraft Contrails." *Physical Review Fluids* 8(11):114702.

Schumann, U. 2012. "A Contrail Cirrus Prediction Model." *Geoscientific Model Development* 5(3):543–580.

Schumann, U., P. Jeßberger, and C. Voigt. 2013. "Contrail Ice Particles in Aircraft Wakes and Their Climatic Importance." *Geophysical Research Letters* 40(11):2867–2872.

Schumann, U., J.E. Penner, Y. Chen, C. Zhou and K. Graf. 2015. "Dehydration Effects from Contrails in a Coupled Contrail–Climate Model." *Atmospheric Chemistry and Physics* 15(19):11179–11199.

Singh, D.K., S. Sanyal, and D.J. Wuebbles. 2024. "Understanding the Role of Contrails and Contrail Cirrus in Climate Change: A Global Perspective." *Atmospheric Chemistry and Physics* 24(16):9219–9262. https://doi.org/10.5194/acp-24-9219-2024.

Sussmann, R., and K.M. Gierens. 1999. "Lidar and Numerical Studies on the Different Evolution of Vortex Pair and Secondary Wake in Young Contrails." *Journal of Geophysical Research: Atmospheres* 104(D2):2131–2142.

Teoh, R., Z. Engberg, U. Schumann, C. Voigt, M. Shapiro, S. Rohs, and M.E.J. Stettler. 2024. "Global Aviation Contrail Climate Effects from 2019 to 2021." *Atmospheric Chemistry and Physics* 24(10):6071–6093.

Thompson, G., C. Scholzen, S. O'Donoghue, M. Haughton, R.L. Jones, A. Durant, and C. Farrington. 2024. "On the Fidelity of High-Resolution Numerical Weather Forecasts of Contrail-Favorable Conditions." *Atmospheric Research* 311:107663.

Unterstrasser, S., K. Gierens, and P. Spichtinger. 2008. "The Evolution of Contrail Microphysics in the Vortex Phase." *Meteorologische Zeitschrift* 17(2):145–156.

Wang, X., K. Wolf, O. Boucher, and N. Bellouin. 2024. "Radiative Effect of Two Contrail Cirrus Outbreaks Over Western Europe Estimated Using Geostationary Satellite Observations and Radiative Transfer Calculations." *Geophysical Research Letters* 51(7).

Wang, Z., L. Bugliaro, T. Jurkat-Witschas, R. Heller, U. Burkhardt, H. Ziereis, G. Dekoutsidis, et al. 2023. "Observations of Microphysical Properties and Radiative Effects of a Contrail Cirrus Outbreak Over the North Atlantic." *Atmospheric Chemistry and Physics* 23(3):1941–1961.

Wolf, K., N. Bellouin, and O. Boucher. 2023. "Long-Term Upper-Troposphere Climatology of Potential Contrail Occurrence Over the Paris Area Derived from Radiosonde Observations." *Atmospheric Chemistry and Physics* 23:287–309.

Wolf, K., N. Bellouin, and O. Boucher. 2024. "Distribution and Morphology of Non-Persistent Contrail and Persistent Contrail Formation Areas in ERA5." *Atmospheric Chemistry and Physics* 24(8):5009–5024.

Zhou, C., and J.E. Penner. 2014. "Aircraft Soot Indirect Effect on Large-Scale Cirrus Clouds: Is the Indirect Forcing by Aircraft Soot Positive or Negative?" *Journal of Geophysical Research: Atmospheres* 119(19):11303–11320.

5

Contrail Forecast and Verification

Modeling systems described in Chapter 4 can and have been used to assess the overall climate impact of contrails (e.g., Lee et al. 2021), as described in Chapter 1. These modeling systems can also be used to forecast contrails. For either use, models require verification against observations at many levels. This chapter will outline the purpose of contrail forecasting systems, a description of current efforts, verification methods using observations, the elements of what a complete forecast system might look like, and the challenges and opportunities involved in building and operationalizing a contrail forecast system.

THE PURPOSE OF A CONTRAIL FORECAST SYSTEM

As previously discussed, the purpose of modeling contrails can be to assess the overall impact of contrails or to forecast specific contrails (or groups of contrails), which may be used in operational contrail avoidance. This involves several steps and inputs, as indicated in Figure 5-1. The structure of the system also depends on the outputs. The output could simply be whether a contrail forms at all (the military has an interest in this). The purpose could be to determine where contrails form and persist in ice-supersaturated regions (ISSRs). It could also include a representation of aviation locations and emissions to estimate the radiative forcing from an individual or set of contrails.

Regardless of the intent, a forecast system requires a series of computational elements described in detail in Chapter 4. First is a model of the atmosphere. To produce a forecast, the model must be initialized to the current state of the atmosphere, bringing in observations to the model with a data assimilation system. The purpose is to use all available observations (e.g., those described in Chapter 3) to describe the current state of the system. This essentially is a weather forecast system, or if run for a long time when the initial conditions do not matter, a climate model and climate projection. Such a model would produce locations of ISSRs at a given future time. This could be used directly, or with aircraft position and emission data, a representation of contrails in another model or a parameterization within the forecast model used. Note that if the plume/contrail representation is embedded in the forecast system, then the contrail would evolve with the model over time. The result of such a plume model or contrail parameterization would be an estimate of contrail lifetime, size (dispersion), and hence, the radiative forcing impact. Density of multiple or overlapping layers of contrails is also important (although it may be rare to fly exactly through the same plume). It may be possible to use data-driven methods trained on observations, models, aircraft positions, and contrail observations to directly calculate contrail effects from the environmental

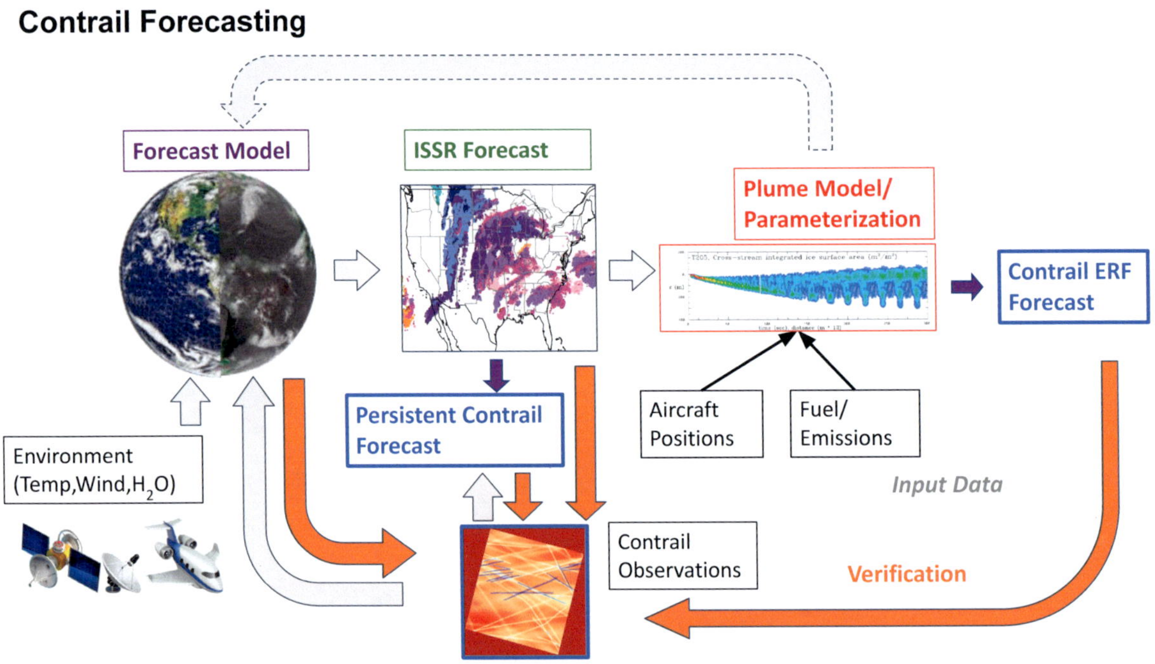

Contrail Forecasting

FIGURE 5-1 Schematic elements of a contrail forecasting and verification system illustrating how a forecast model (purple) provides an ice-supersaturated region (ISSR) forecast (green) which then can be used with a contrail plume model or parameterization (red) to estimate the contrail effective radiative forcing (ERF) (blue). Contrail observations can be used to verify the predictions (as well as one of the inputs to the system). Orange arrows represent verification, where the different pieces of the system can be verified against observations of contrails.
SOURCES: (Earth and ISSR Forecast) Courtesy of NASA; (Plume Model/Parameterization) Used with permission of D.C. Lewellen, O. Meza, and W.W. Huebsch, 2014, "Persistent Contrails and Contrail Cirrus. Part I: Large-Eddy Simulations from Inception to Demise," *Journal of the Atmospheric Sciences* 71(12):4399–4419; permission conveyed through Copyright Clearance Center, Inc.

properties. There are some efforts in this regard, but they also require use of similar observations (Chapter 3) and models (Chapter 4) for training.

Critical to such a system is evaluation or verification. Verification is a way of assessing and continuously refining forecasts checking against observations. This means that atmospheric and contrail observations need to be used a posteriori to evaluate forecast accuracy. Models are typically "evaluated" for their performance, while specific predictions are "verified" with trusted observations. In the case of a contrail forecasting system (also for a climate model with contrails), the forecasts can be evaluated in key ways.

First, observations of the atmosphere at the forecast time can be evaluated to see if ice supersaturation was predicted. Second, observations of contrails (e.g., from satellite or ground-based systems) can also be used to evaluate the forecast. This information ideally would feed back to improve the model, and/or determine where data gaps are most important. Currently, individual tracks are difficult to verify, due to a dearth of observations. Observations of temperature and humidity (Chapter 3) will be used for designing contrail forecast systems (Chapter 4) and verifying them. The observations from commercial aircraft of temperature, humidity, and any visible contrails are the ultimate evaluation of model skill. These are described in more detail in Chapter 3 and later in this chapter.

Note that the same methodology could be used to describe and evaluate a system to estimate the aggregate radiative effect of contrails. Detailed verification/evaluation of a forecast system could enable "hindcasts" to estimate aggregate effective radiative forcing (ERF), and has been used for such evaluations (e.g., Teoh et al. 2024).

The system could use a climate model with a contrail parameterization (e.g., Bock and Burkhardt 2016; Gettelman et al. 2021), or a plume model run offline (e.g., Teoh et al. 2024). Such estimates also can be "verified" against observations, often statistically (do they produce the right statistics of ISSRs, for example). The main challenge is having confidence in representing contrail cirrus clouds, which are harder to evaluate.

EXISTING CONTRAIL FORECAST SYSTEMS

Forecasting, nowcasting, and hindcasting of contrails and contrail-forming regions (ISSRs) has been undertaken for decades. The original motivation of such systems was related to military purposes to predict any contrail (not just persistent contrails) that might aid ground-based location of aircraft. Over the past two decades, research has used hindcast meteorological products that include parameters necessary to estimate ice supersaturation. More recently, observationally based nowcasting has been attempted using numerical weather prediction models. These efforts have revealed the difficulty of obtaining reliable ISSR nowcasts/forecasts from current systems discussed in Chapter 4.

ICE-SUPERSATURATED REGION NOWCASTING/FORECASTING

The U.S. Air Force and other armed forces around the world have long practiced contrail management. This has been motivated by the need to reduce detection or, in some cases, the objective of contrail avoidance. The U.S. Air Force produces contrail forecasts for its own use (AWS 1981; Peters 1993). However, where national security is not compromised, there may be the opportunity for the defense community to share learning with the climate science and aviation communities.

Recently, machine learning–based approaches to identifying contrails visible from geostationary satellites have been developed. There are now several research groups (including in the private sector) that are producing automated contrail identification from satellite imagery. This typically entails the development of a deep learning convolutional neural network architecture suitable for the task, combined with the creation of training and verification data sets. Training data sets are based on human-labeled contrails. Furthermore, methods of inferring the current ISSRs from observations of contrails, non-observation of contrails where there are known flights, and length-scale characteristics of ISSRs have been performed. This "estimation kernel" approach has the benefit of being based on observations, but forecasts are limited to extrapolations on the order of 30 minutes or less. Such approaches may be useful for tactical contrail avoidance, or adjusting flight plans that have been filed based on numerical weather prediction systems.

There are nascent ISSR-prediction systems based on numerical weather prediction models as well. The private-sector research company SATAVIA developed a numerical weather prediction tool for ISSR prediction based on a modified version of a numerical weather prediction model (Thompson et al. 2024). Meteo-France has an ISSR prediction model in development, and DWD (German meteorological service) is improving their forecast model to permit and predict ISSRs.

The skill of existing numerical weather prediction systems in predicting space-time-paired ISSRs has not been proven sufficient for operational implementation of navigational contrail avoidance, but may be sufficient for trials. Observational approaches using satellites are currently limited by the relatively low resolution of geostationary satellites. Weather model forecasts of ISSR regions are limited by (1) lack of flight-level in situ humidity data for constraining ISSR locations at flight altitudes, (2) poor parameterizations of cloud physics in weather models that do not permit ice supersaturation or represent ice nucleation, and (3) low vertical resolution of the models in the upper troposphere that cannot reproduce thin layered structures of ISSRs. Progress will depend on advancing weather forecasting codes for this purpose (e.g., using methods developed for global climate models, which do allow ice supersaturation and treat aerosol effects, with high vertical resolution in the upper troposphere), improvement in observations, and assimilation of those observations into forecast systems.

Finding: Current weather forecast systems are not sufficient for operational prediction of ISSRs.

CONTRAIL FORECASTING

Contrail forecasting is strongly contingent on the ability to forecast ISSRs, both in their existence and the extent of ice supersaturation. Forecasts currently have been based on contrail plume models coupled to weather forecast models, or simple heuristics with the Schmidt–Appleman criterion indicating regions of potential persistent contrail formation from weather forecast models. Several examples include the following:

- The U.S. Air Force produces an internal contrail formation product for its own internal use based on weather forecasts and a simple algorithm based on the Schmidt–Appleman criterion.
- The NASA SatCORPS group (NASA Langley) produces a persistent contrail-formation potential product based on geostationary satellite imagery, as well as contrail forecasts based on operational weather model forecast output (the Rapid Refresh and North American Mesoscale models from the National Oceanic and Atmospheric Administration [NOAA]).[1]
- The German Aerospace Center (DLR) produces internal use forecasts of Schmidt–Appleman criterion temperature difference (like a dew point) and ice supersaturation computed from European Centre for Medium-Range Weather Forecasts (ECMWF) Integrated Forecasting System (IFS) in support of operational flight campaigns. Information from contrail cirrus prediction model (CoCiP) simulations also provides forecasted contrail formation altitudes and contrail optical depth.
- Breakthrough Energy has created an organization, called Contrails.org, which publishes "real time" contrail coverage estimates[2]). These are based on using weather forecasts to initialize the CoCiP contrail plume model discussed in Chapter 4.

Contrail forecasting accuracy is most strongly affected by the ISSR forecasts—both existence and extent of ice supersaturation. To date, models for contrail forecasting have been repurposed models intended for retroactive analyses. At this stage, it is unlikely that individual ERF estimates made in forecast are accurate, whereas averages over large numbers of contrails have lower uncertainty because averages are better constrained by sampling across weather and aircraft variability from the microscale (contrail ice particle number) to the global scale (average frequency of ISSRs).

Since these methods are dependent on ISSR forecasts, they suffer from the same problems as current predictions of ISSRs—lack of in situ humidity data and the inability of forecast model resolution and cloud parameterizations to represent ISSRs. Furthermore, because ambient aerosols become more important with reduced soot, to estimate the potential future impact of contrail evolution and contrail cirrus, models would need to represent ambient aerosol and ice nucleation. This is done with much uncertainty in one of the general circulation models (Gettelman and Chen 2013) but is not present in any weather forecast system.

Finding: Current forecasts of ISSRs and persistent contrails are limited by (1) lack of flight-level in situ humidity data for constraining ISSR locations at flight altitudes, (2) poor parameterizations of cloud physics in weather forecast models that do not permit ice supersaturation, (3) low vertical resolution in the upper troposphere, and (4) no representation of ice nucleation linked to particulates (aerosols).

NASA already has a modeling system (the Goddard Earth Observing System [GEOS]) that has a comprehensive data assimilation system that can ingest humidity and temperature data from aircraft if available, along with current satellite observations, and has a good representation of ice supersaturation and ice nucleation. Furthermore, the Global Modeling and Assimilation Office has expertise in ice nucleation and the physics of the upper troposphere. This system could be used as a prototype for a contrail evaluation and forecast system. In addition, NASA has an observation system simulation experiment (OSSE) framework available to assess the utility of new observations on producing forecasts to determine if and where new observations could improve forecasts.

---

[1] NASA, "Contrails," Satellite Cloud and Radiation Property Retrieval System (SatCORPS) Group, https://www-pm.larc.nasa.gov/cgi-bin/site/showdoc?mnemonic=CONTRAIL_FORECAST.

[2] Contrails.org, "Experience How Contrails Impact the Climate," https://map.contrails.org, accessed December 1, 2024.

Finding: Reliable prediction of ISSRs is likely possible and feasible with evolution of current models and sufficient humidity and temperature data at flight level.

Recommendation: NASA should apply its current Earth system modeling efforts in support of simulating ice-supersaturated regions and contrails as a pathway to demonstrate the use of observations and advanced modeling tools for developing a contrail forecast and prediction system and estimating contrail radiative forcing.

EVALUATION METHODS

Evaluation of ISSR reanalyses/forecasts/nowcasts and contrail forecasts is necessary to quantify the data that may be used for contrail avoidance or any other approaches to contrail mitigation.

Much of the direct evaluation of ISSRs has focused on reanalyses. Gierens et al. (2020) first highlighted challenges in accurately estimating ISSRs using existing numerical weather prediction models with reference to airborne measurements from MOZAIC flights. This was followed up by Hofer et al. (2024), who used ERA5 predictions, finding that ISSR prediction is "very difficult." Agarwal et al. (2022) evaluated reanalysis products ERA5 and MERRA-2 from almost 800,000 radiosondes. Results were interpreted as showing that persistent contrail formation at cruise altitudes is overestimated by a factor of 2–3.5 when using these products, and that a contrail lifetime metric is overestimated by 17–45 percent. They also estimate that the existing products incorrectly identify the regions of contrail formation and persistence 52–87 percent of the time. Thompson et al. (2024) also evaluated two operational forecasts (ECMWF IFS and NOAA Global Forecast System [GFS]) with their specialized model and found no skill in the GFS for predicting contrail formation, and only moderate skill for IFS or the specialized SATAVIA version of the weather research and forecasting model. Overall, the evaluations suggest that weather models alone in their current state of development are unlikely to be able to support operational contrail avoidance due to inaccuracies (operational aspects are discussed in Chapter 6).

Finding: Current attempts at operational forecasting of contrail locations and their radiative effects have low predictive skill, making their use in climate mitigation metrics questionable.

Additional scope for evaluation and potentially operation or real-time forecasting lies in satellite-based contrail observations. Furthermore, observations of contrails are critical for validation and verification of avoidance. This work was largely pioneered by Minnis et al. (2003). Systems such as GOES in mesoscale mode provide 500-m resolution with several minutes time resolution, sufficient for tracking individual contrails. NASA SatCORPs has also been archiving satellite data over North America from which to do contrail detection. New machine learning methods can automate detection of contrails as developed by Meijer et al. (2022), and attribution of contrail evolution to images over time (Hoffman et al. 2023), and data sets (e.g., OpenContrail; Ng et al. 2024) now exist to facilitate this work.

Finding: New machine learning methods for contrail identification from satellite imagery are showing great promise, but databases are currently fragmented, use different methods, and are not global.

Specialized research satellites with multi-spectral observations, such as NASA's MODIS satellites, now transitioned to operational systems in low Earth orbit (LEO) (such as VIIRS), as well as new multi-spectral imagery from geostationary orbits (GEOs) soon to be available (see Chapter 3), can provide not just contrail locations (including some vertical information from brightness temperature), but some information on contrail microphysics (optical depth, and even some information on particle size). Furthermore, broadband radiation sensors (e.g., CERES and follow-ons) can provide direct measurements of individual and multiple contrail (scene-level) radiative forcing.

Finding: LEO and geostationary satellite data can be used with aviation location data and machine learning to improve model accuracy and is vital for model validation and testing.

As discussed in Chapter 3, the use of ground-based cameras has been proposed and early work has been done by DLR, the Massachusetts Institute of Technology (MIT), Cambridge University, Imperial College, and other research organizations. This may be useful for flights over land where camera networks could be created to supplement satellite or other observations. However, the utility of ground-based cameras is limited by the need for clear skies. Cameras are also useful for validating satellite-based instruments and can provide vertical information if matched to flight data. A consistent approach to evaluation and standard metrics for contrail observation and identification has not been developed.

Finding: Verification of contrail predictions requires good observations of contrails globally. This requires use of multiple satellites and ground-based cameras, in an open-source setting.

Recommendation: NASA should support development of a global contrail-observing system as a foundation for research, analysis, and future verification.

There has been discussion by private actors about building a specialized satellite or constellation of satellites for global observations of contrails. A purpose-built system could take the best aspects of low Earth orbit sensors for imaging and radiation with some humidity estimation, and advance contrail identification and potentially prediction (see below). NASA has valuable expertise in Earth observations for instruments and data processing algorithms (retrievals) that could assist with optimal development of such a system, and ensure public use of any data. This could massively advance evaluation (and possible prediction) of contrails. Should an international mitigation plan for contrails be put in place for verification, or should others decide to build a dedicated satellite system for contrail observation and prediction, NASA can collaborate to ensure quality science and open data access for any contrail observations.

Finding: Non-governmental and private-sector parties are exploring the deployment of a specific constellation of satellites to observe contrails. NASA is in a unique position to advise these efforts to ensure that they are best architected and that resulting data are accessible to maximize utility for the broader community.

WHAT WOULD A FORECAST SYSTEM LOOK LIKE?

Existing systems do not have all the elements of a forecast system illustrated in Figure 5-1, but systems could be of different types.

Perhaps the most obvious system uses a weather forecast approach, where a physical model produces a forecast of the state of the atmosphere, and then that state can be used to calculate ice supersaturation and further used as input to a model of aircraft contrails. Note that a weather prediction model and a climate model often contain the same physical representations, but a climate model is integrated sufficiently far from the initial conditions and with random variability introduced that it no longer is intended to reproduce the exact same final state from a set of initial conditions on each run. Weather models are also used as hindcasts in "analysis" mode: the data assimilation system uses observations to generate the most consistent possible state of the atmosphere, and this is saved for each time, generating a "re-analysis" that can be used as a uniform set of atmospheric conditions in the past (our "best estimate" of the system based on all available observations). Weather forecast models can be both regional and global. A typical global weather forecast model has horizontal resolution on the order of 10–20 km, while regional models are nearly an order of magnitude finer (1–5 km).

Current operational weather forecasting systems are not specifically designed to forecast ice supersaturation or contrails. They are designed for surface weather (temperature, precipitation) and many processes in the upper troposphere are not sufficiently represented in most weather models. This is also true of re-analysis systems used as "hindcasts" for assessing the overall contrail ERF. In addition, sufficient observations at flight level, particularly of humidity, are needed to be able to assimilate into weather models to get more accurate initial states. Very few models have representations of ice supersaturation; rather they assume ice clouds form at 100 percent relative humidity over ice (or slightly earlier to account for sub-grid variability). This makes it difficult to use current forecast or reanalysis systems for ISSR and contrail forecasting. Climate models used for contrail estimates (e.g., Bock and Burkhardt 2019; Chen and Gettelman 2013) have more advanced representations of clouds and humidity that do represent ISSRs (e.g., Gettelman et al. 2010). There are now some advancements putting such representations into weather models, such as efforts in Germany to add ice supersaturation to their global weather model. A forecast of ice supersaturation with the Schmidt–Appleman criterion (sufficiently cold to form a contrail) would then indicate where persistent contrails would form (as is being attempted now by the U.S. Air Force, NASA SatCORPS, and DLR), or enable use of a plume parameterization/model (such as CoCiP used by pycontrails[3]).

In addition to a lack of representation of ice supersaturation, the lack of resolution in a forecast model is also a substantial problem. The horizontal resolution is not the major factor: most models have the ability to represent partial cloudiness generated by contrails, and the atmosphere in the upper troposphere tends to be well mixed in the horizontal on scales of several to tens of kilometers. However, the upper troposphere is very stratified in the vertical, and observations indicate ice supersaturation layers are several hundred meters thick (Spichtinger et al. [2003] found over Germany the average layer was 560 m with a standard deviation of 610 m), while a typical horizontal extent is 150 km (Gierens and Spichtinger 2000). Most weather and climate models have vertical resolution of 1 km or larger in the upper troposphere. An ideal contrail forecasting system should also have high vertical resolution (200–300 m or less) to resolve thin ISSRs.

Finding: A contrail prediction system requires a forecast model with representation of ice supersaturation, which necessitates accounting for ice supersaturation and higher vertical resolution.

Another approach to estimating ice supersaturation and related to verification is a "nowcast" approach: using satellite and ground-based observations of current contrails along with known aircraft altitudes to determine where ice-supersaturation regions are in flight lanes at the current time. It can also tell where there is subsaturation if aircraft are not forming contrails. This information can be propagated forward in time with forecasts of wind from weather models, or used directly. For example, contrail observations have been used to trial operational avoidance in experiments done by MIT and Delta Air Lines, where observations of contrails and known aircraft locations without contrails are being used to infer contrail-forming regions. Such an approach could not eliminate 100 percent of contrails as some fraction are needed as "tracer aircraft." Observations of persistent contrails as noted above are needed for forecast verification in any event. New satellite products coming online with machine learning methods for automated detection of contrails (discussed in Chapter 3) would be useful in this regard.

As described in Chapter 4, the next step in such a system to estimate contrail radiative effects (either globally for climate estimates or for an individual contrail) would be to have a representation of contrails in the modeling system or to use a forecast (or analysis) to drive a contrail model. This requires knowledge of aircraft locations and an estimate of aircraft emissions (usually based on aircraft and engine type). The representation could be a simplified plume model—for example, CoCiP or APCEM (see Chapter 4), a simplified representation of ice cloud formation from aircraft (e.g., Bock and Burkhardt 2019; Chen and Gettelman 2013), or even just a rough estimate or assumption of the radiative effect of a cloud given the environmental conditions. On an aircraft-by-aircraft basis, contrails can be "nowcast" from the aircraft, with temperature and humidity sensors to measure the relative humidity and temperature in situ and/or with a camera to try to visualize behind the aircraft, or following other aircraft.

What input data would be needed for such systems? Currently weather forecast information comes from a wealth of sensors such as ground-based data, satellite observations, and profiles from weather balloons. In addition, as described in Chapter 3, wind and temperature data from aircraft are assimilated into forecast systems to improve the winds at flight level. This has been successful and valuable for aviation. Limited water vapor sensors currently on aircraft are also assimilated now, mostly on a profile basis (at climb and descent) to improve the profiling of water vapor in the atmosphere. These profiles have been shown to improve icing forecasts at lower altitudes. It is straightforward to assimilate aircraft flight-level data if new systems are added as discussed in Chapter 3. As with improvements in flight-level wind due to in situ observations, the addition of water vapor data to assimilation systems used in forecasting would dramatically improve ISSR forecasts, at least in the short term of 24 hours

---

[3] See the pycontrails website at https://py.contrails.org, accessed December 1, 2024.

or so. Note that a simple analysis of the Schmidt–Appleman criterion and temperatures at aircraft flight altitudes indicates that at typical flight levels (30,000–45,000 feet), any water vapor concentration below about 20 ppmv is too dry to support contrail formation or ice supersaturation. This is much higher than the values of water vapor in the stratosphere, but it is simply sufficient to know that water vapor is less than 20 ppmv (see Chapter 3).

In addition, satellite information on contrail locations from current or future sensors could be used to locate persistent contrails. If the information were available in near real time, this is an indication of ice supersaturation (qualitatively) that could be used in modeling. It is also vital for verification of contrail forecasts. Ground-based cameras could be used in a similar manner, but would only be able to sense in clear sky conditions, while space-based systems could use infrared bands to discriminate between high and cold contrails, and lower, warmer clouds. Space- or ground-based "cameras" would ideally need information on aircraft altitudes to be able to locate ISSRs. Cameras on planes could also indicate contrail formation (or lack thereof) but not persistence. It is clear that a rear-facing camera on a wing or the tail would not be able to see a persistent contrail behind an aircraft, given aircraft speed. If an aircraft is traveling 900 km/hr, then it travels 15 km a minute and would not be able to see a persistent contrail behind it.

> Finding: New observations for humidity and even contrail imagery would improve forecasting ISSRs and contrails.

An "ideal" contrail forecasting system structure would depend on the desired goal of the system. Most goals require forecasting ISSR locations. This is sufficient to determine regions of avoidance at the regional or individual flight level. The uncertainty on ISSR forecasting for larger regions, especially with the right input data and model, is tractable and reducible. The most critical elements noted above are probably in situ humidity data from aircraft assimilated into models that permit ice supersaturation and have relatively fine vertical resolution at flight altitudes. Persistent contrail observations with a consistent global system would also be useful for assimilating the current locations of ISSRs in flight regions and are vital for verification.

This is a prerequisite for forecasts of individual contrails. However, individual contrails and their radiative forcing are subject to a high degree of uncertainty due to smaller scale variations in the atmosphere, and the details of fuel and engine emissions characteristics (see Chapter 2). Estimating ice mass, optical depth, and radiative forcing are highly uncertain and would require more information (like the quantitative level of ISSR). Much less uncertainty would be attached to predictions of large ISSRs that intersect flight locations.

Note that predicting the evolution and fate of aviation-induced cloudiness (clouds affected by aviation particulate emissions not directly behind an aircraft) is even more difficult, and probably only tractable to do in a statistical or hindcast sense. These models must include detailed ice-nucleation and aerosol representations (e.g., Chen and Gettelman 2013). There are still large uncertainties in aviation-induced cloudiness (not even assessed in Lee et al. 2021), which need to be addressed with continued development and research using observations and models (likely a side effect of developing an observational and verification system for contrails). There are also important uncertainties in how alternative fuels that create different populations of particulate matter (discussed in Chapter 2) will alter the climate impact of contrails.

> Finding: Understanding and predicting aviation-induced cloudiness requires global models designed to simulate aerosol evolution and ice nucleation, evaluated with measurements of aerosols.

CHALLENGES AND OPPORTUNITIES

This chapter has illustrated a roadmap for taking the observations described in Chapter 3 and models described in Chapter 4 and building a system designed to simulate where contrails will form, and their radiative effect. The systems would have two purposes: predicting contrails for mitigation purposes (and verification), as well as continued research into the fundamental physics of contrail effects to reduce uncertainties particularly in aviation-induced cloudiness and particulate effects that are relevant to future aviation engine emissions from new technologies and alternative fuels. This roadmap is still rough since the final objectives of a system need to be defined.

Forecasting of ISSRs and contrails with current weather forecast and contrail models is feasible and shows promise (e.g., Thompson et al. 2024). However, quantitative skill scores (mostly of ISSR predictions) are not high enough to reduce uncertainty in contrail impacts or to do reliable contrail avoidance. Furthermore, observations of contrails are not sufficient to verify contrail formation regions or to further constrain uncertainties in models, particularly in contrail plume models.

One need is to improve models, particularly for simulating ISSRs and the representation of particulate matter effects on clouds. Improvement of ice supersaturation and ice nucleation is needed. Models will also need higher vertical resolution in the upper troposphere than current global forecast systems. Note that this either requires engagement with weather forecast services, or duplicating their work in a specific system for forecasting contrails in the upper troposphere.

A second need is to improve observations used for (1) forecasting and (2) verification. For forecasting, there are good flight-level temperature data, but not sufficient flight-level humidity data, which likely will have to come from in situ sensors to get the vertically resolved information necessary. Satellite-based information on environmental state (temperature and humidity) is expected to improve over time but will not be sufficient on its own, because the vertical resolution of satellite instruments will never be able to sufficiently resolve the vertical gradient in water vapor.

Verification of contrails or avoidance requires observations of contrails. This is most usefully done by satellites and ground-based cameras. New machine learning algorithms are showing great promise in automated detection of contrails. Current satellite systems are useful for this, and future systems will be better. Ground-based imagery may be able to supplement and help evaluate/verify such estimates. Cameras on aircraft may also provide some information on instantaneous (but not persistent) contrails. This work may be advanced with a specialized constellation of small satellites to observe contrails, and some efforts are being discussed in this area. In any case, better coordination across current activities is needed to provide an open and evaluated global database of contrails (presence, and even optical and radiative properties). This is possible but would require focused investment and attention.

There are many uncertainties in contrail prediction, but forecasting of large ISSRs and prediction of the radiative impact of aviation emissions in those regions is likely possible to enable effective contrail avoidance. Verification of such predictions is also possible with appropriate observations. Uncertainty of contrail predictions for individual flights is very large, and verification of individual flight radiative effects is extremely difficult, and to some extent irreducible. Verification is critical for any individual flight mitigation and has a higher bar than verification of ISSR predictions in large regions with contrails. Uncertainty should not prevent the development of observations and forecast systems for contrail prediction.

There are several critical priorities to enable contrail mitigation in the near term with higher certainty. First are high-vertical-resolution forecasts, which require improving weather forecast models. Second is estimating high-RHi regions and ISSRs, which requires better in situ humidity observation, probably from commercial aircraft.

Finding: Highest priorities for enabling contrail mitigation systems are high-vertical-resolution forecasts of high RHi and ISSRs, which would likely require better forecast models that permit ice supersaturation and better in situ humidity observations.

To operationalize a contrail mitigation system, it would be best to focus on more certain contrails with known high impact. High-impact contrails have large positive radio frequency, which comes from long lifetime, high RHi, and certain conditions, like nighttime over warm surfaces. Development of contrail prediction systems could help refine high-impact contrail locations. For impact before the end of this decade, activities should focus on what can be done in the current air traffic control system, which likely means moderate vertical rerouting.

Chapter 6 will focus on operational concepts that can enable contrail mitigation within the aviation system.

REFERENCES

Agarwal, A., V.R. Meijer, S.D. Eastham, R.L. Speth, and S.R.H. Barrett. 2022. "Reanalysis-Driven Simulations May Overestimate Persistent Contrail Formation by 100%–250%." *Environmental Research Letters* 17:14–45.

AWS (Air Weather Service). 1981. *Forecasting Aircraft Condensation Trails.* Technical Report AWS/TR-81/001. U.S. Air Force.

Bock, L., and U. Burkhardt. 2016. "The Temporal Evolution of a Long-Lived Contrail Cirrus Cluster: Simulations with a Global Climate Model." *Journal of Geophysical Research: Atmospheres* 121(7):3548–3565.

Bock, L., and U. Burkhardt. 2019. "Contrail Cirrus Radiative Forcing for Future Air Traffic." *Atmospheric Chemistry and Physics* 19(12):8163–8174.

Chen, C.-C., and A. Gettelman. 2013. "Simulated Radiative Forcing from Contrails and Contrail Cirrus." *Atmospheric Chemistry and Physics* 13(24):12525–12536.

Gettelman, A., and C-C. Chen. 2013. "The Climate Impact of Aviation Aerosols." *Geophysical Research Letters* 40(11):2785–2789. https://doi.org/10.1002/grl.50520.

Gettelman, A., X. Liu, S.J. Ghan, H. Morrison, S. Park, A.J. Conley, S.A. Klein, J. Boyle, D.L. Mitchell, and J.-L.F. Li. 2010. "Global Simulations of Ice Nucleation and Ice Supersaturation with an Improved Cloud Scheme in the Community Atmosphere Model." *Journal of Geophysical Research: Atmospheres* 115(D18).

Gettelman, A., C.-C. Chen, and C.G. Bardeen. 2021. "The Climate Impact of COVID-19-Induced Contrail Changes." *Atmospheric Chemistry and Physics* 21(12):9405–9416.

Gierens, K., and P. Spichtinger. 2000. "On the Size Distribution of Ice Supersaturation Regions in the Upper Troposphere and Lowermost Stratosphere." *Annales Geophysicae* 18:499–504.

Gierens, K., S. Matthes, and S. Rohs. 2020. "How Well Can Persistent Contrails Be Predicted?" *Aerospace* 7(12):169.

Hofer, S., K. Gierens, and S. Rohs. 2024. "How Well Can Persistent Contrails Be Predicted? An Update." *Atmospheric Chemistry and Physics* 24(13):7911–7925.

Hoffman, J.P., T.F. Rahmes, A.J. Wimmers, and W.F. Feltz. 2023. "The Application of a Convolutional Neural Network for the Detection of Contrails in Satellite Imagery." *Remote Sensing* 15(11):2854. https://doi.org/10.3390/rs15112854.

Lee, D.S., D.W. Fahey, A. Skowron, M.R. Allen, U. Burkhardt, Q. Chen, S.J. Doherty, et al. 2021. "The Contribution of Global Aviation to Anthropogenic Climate Forcing for 2000 to 2018." *Atmospheric Environment* 244:117834.

Meijer, V.R., L. Kulik, S.D. Eastham, F. Allroggen, R.L. Speth, S. Karaman, and S.R.H. Barrett. 2022. "Contrail Coverage Over the United States Before and During the COVID-19 Pandemic." *Environmental Research Letters* 17(3):034039. https://doi.org/10.1088/1748-9326/ac26f0.

Minnis, P. 2003. "Contrails." Pp. 509–520 in *Encyclopedia of Atmospheric Sciences* (J. Holton, J. Pyle, and J. Curry, eds.). Academic Press.

Ng, J.Y.H., K. McCloskey, J. Cui, V.R. Meijer, E. Brand, A. Sarna, N. Goyal, C. Van Arsdale, and S. Geraedts. 2024. "Contrail Detection on GOES-16 ABI with the Opencontrails Dataset." *IEEE Transactions on Geoscience and Remote Sensing* 62:1–14.

Peters, J.L. 1993. *New Techniques for Contrail Forecasting.* Technical Report AWS/TR-93/001. U.S. Air Force Weather Service.

Spichtinger, P., K. Gierens, U. Leiterer, and H. Dier. 2003. "Ice Supersaturation in the Tropopause Region Over Lindenberg, Germany." *Meteorologische Zeitschrift* 12(3):143–156.

Teoh, R., Z. Engberg, U. Schumann, C. Voigt, M. Shapiro, S. Rohs, and M.E.J. Stettler. 2024. "Global Aviation Contrail Climate Effects from 2019 to 2021." *Atmospheric Chemistry and Physics* 24(10):6071–6093.

Thompson, G., C. Scholzen, S. O'Donoghue, M. Haughton, R.L. Jones, A. Durant, and C. Farrington. 2024. "On the Fidelity of High-Resolution Numerical Weather Forecasts of Contrail-Favorable Conditions." *Atmospheric Research* 311:107663.

6

Operational Concepts

Assuming that the radiative forcing impact of aviation-induced contrails is sufficiently strong to justify mitigation measures, it is likely that operational strategies that avoid flying through regions that would generate the most impactful warming contrails would be the most effective mitigation measure (Kärcher 2018). This is particularly true in the near term as operational contrail avoidance can be done with the existing aircraft fleet as other mitigations such as low-emission alternate fuels and improved combustion technology will take time to develop, certify, and be implemented at large scale.

The industry has demonstrated interest and willingness to participate in operational avoidance as demonstrated in the recent operational avoidance trials shown in Box 6-1. These trials demonstrate the capability of operators to modify routes and altitudes but are not capable of validating the effectiveness of contrail avoidance because there are currently insufficient forecast skills to know if the unmodified routes would have generated persistent contrails. In addition, the limited number of flights do not address the air traffic control (ATC) issues that would emerge if significant numbers of flights were to request contrail avoiding routes and altitudes.

There is emerging government interest in operational contrail mitigations, particularly in the European Union, which has recently imposed non-CO_2 emission monitoring, reporting, and verification requirements on air carriers which operate within Europe.

> Finding: The global policy landscape is evolving rapidly and is trending toward more widespread monitoring and control of non-CO_2 aviation impacts, which may result in the adoption of operational avoidance efforts over the coming decades.

Any operational contrail avoidance approach must balance the total change in radiative impact due to the reduction of contrails against any potential additional CO_2 or other emissions generated by avoidance trajectory. Because of the difference in impact timescales and uncertainty in contrail radiative forcing, the exact trade-off between contrail and additional CO_2 is not fully known, but there are clearly some situations where high-impact potential contrails could be avoided with limited additional CO_2 emissions and would have a net positive environmental impact.

> Finding: Some specific contrails contribute notably to climate warming, and mitigating them through operational measures in situations that do not introduce significant additional CO_2 emissions is a net positive environmental strategy.

> **BOX 6-1**
> **Some Recent Operational Avoidance Trials**
>
> **Air France, Meteo France:** Meteo France provided predictions of contrail-prone areas (ice-supersaturated regions [ISSRs]) and Air France undertook operational trails avoiding those regions. The number of flights were limited and a diversion was not always possible due to airspace constraints. The partnership is now working on a campaign for pilots to take photographs of contrails at altitude to verify and validate Meteo France's weather predictions.
>
> **Delta Air Lines, Massachusetts Institute of Technology (MIT):** Delta conducted a contrail avoidance trial with certain flights and MIT used satellite observations of flights and contrails and attempted to verify whether current weather models provided successful reroutings. Results from the trial have not yet been published in peer-reviewed scientific journals.
>
> **American Airlines, Google Research, Breakthrough Energy:** Limited flight trials (around 70 flights) were conducted by American Airlines to determine contrail formation from certain flights using satellite observation of flights and contrails and artificial intelligence in "almost" real time. Questions remain as to the methodology and usefulness of results due to difficulties in validating contrail avoidance, and uncertainties in the model and satellite observations.
>
> **Eurocontrol Maastricht Upper Air Centre (MUAC):** Overnight contrail avoidance trial. The study found challenges in determining the limits of horizontal bands of ISSRs. Tactical simulations with controllers discovered some safety challenges and a potential 20 percent capacity hit.
>
> **Airbus, Meteo France, IAGOS, DLR, NLR, ONERA, UPC, Imperial College, Breakthrough Energy, Air France, Swiss International Airlines, easyJet, NATS, DSNA, Eurocontrol, Boeing:** The SESAR-funded CICONIA project (2023–2026) develops mitigation Concepts of Operations and their assessment in comparison to legacy operations. CICONIA integrates CO_2 and non-CO_2 trade-offs, metrics, the integration of different climate models and aircraft specificities. Extensive simulations and trials in oceanic and continental airspace will be performed.
>
> SOURCE: "Waypoint 2050," Air Transportation Action Group, Briefing Paper #20, June 2024.

Finding: Even while the positive warming impact of some contrails is clear, continued research to reduce uncertainties on climate impacts and traceability is needed to incentivize airlines and regulators to introduce the additional cost and operational complexity associated with introducing broad avoidance measures.

The geometry of contrail-forming ice-supersaturated regions (ISSRs) will strongly impact the feasibility of operational contrail avoidance. Many observational studies have documented that contrail-forming ISSRs can have significant horizontal extent, and most horizontal deviations from a wind-optimal flight route to avoid contrail formation are likely to have a net negative impact on radiative forcing as the fuel burn and resulting CO_2 emission will scale with the additional flight distance.

In contrast, altitude deviations for contrail avoidance hold more promise—because ISSRs are thought to have vertically stratified structure, fuel efficiency for transport aircraft scales weakly for small changes in cruise altitude, and aircraft commonly fly off the fuel-optimal cruise altitudes to conform with standard flight levels and ATC restrictions. A deviation of 2,000 ft would result in less than a 1 percent fuel burn increase for a typical narrow body aircraft and less than an 0.5 percent increase for a typical wide body aircraft (Jensen and Hansman 2014). Since the altitude deviation would only be required for the contrail-impacted duration of the flight, vertical deviations are likely to be a viable avoidance strategy, depending on net value of avoiding the contrail.

Finding: Evaluating the feasibility of effective (net-positive) operational contrail avoidance is dependent on understanding the vertical structure of contrail-forming ISSRs.

Assuming a future environment where the magnitude of contrail impact has been determined to be sufficient to justify policy incentives for operators and Air Navigation Service Providers (ANSPs) to reduce high-impact contrails, an information flow for a general concept of operations for operational avoidance can be seen in Figure 6-1, highlighting the required elements for operation contrail avoidance.

The first required elements are high-quality and high-vertical-resolution forecasts of the contrail-forming ISSRs, as well as a forecast of the radiative forcing (RF)/effective radiative forcing impact of any formed contrail that would include contrail persistence, optical depth, time of day, and natural cloud layers, albedo, and surface information. Current forecast models are insufficient (particularly in the vertical dimension) to support operational contrail avoidance. These forecast models could be improved and informed by any available contrail-monitoring technologies (remote or in situ). This requires observations discussed in Chapter 3, models in Chapter 4 integrated into forecast systems for ISSRs, and warming contrails in Chapter 5.

Finding: Improving the ability to forecast ISSRs is critical for operational mitigation of warming contrails.

The ISSR and contrail impact forecasts would be used by airline operators in their flight planning processes that file requested routes, altitudes, and departure times with ATC (i.e., ANSP). It should be noted that the vertical structure of ISSR regions will be critical to the effectiveness of operational mitigations because ISSRs are thought to be vertically limited and stratified. If this is found to be true, small changes in requested flight levels could mitigate contrails with minor impacts in fuel burn and CO_2 emission. The airline-requested flight plans would be optimized to minimize total operational costs, which would include fuel burn, schedule, turbulence, and any contrail-avoidance incentives.

The requested flight plans are evaluated and modified by ATC to account for safety (aircraft separation, weather, other factors), efficiency, and the capacity of the airspace and airports. ATC issues route clearances that authorize aircraft to fly a specific route and altitude. These routes and altitudes can be modified by the tactical air traffic controllers at the request of the aircraft. This is often done for ride quality reasons due to turbulence but could also be done for unpredicted contrail generation. Depending on future policies, ATC could also have a role in contrail avoidance by limiting access to high contrail-impact airspace or flight levels.

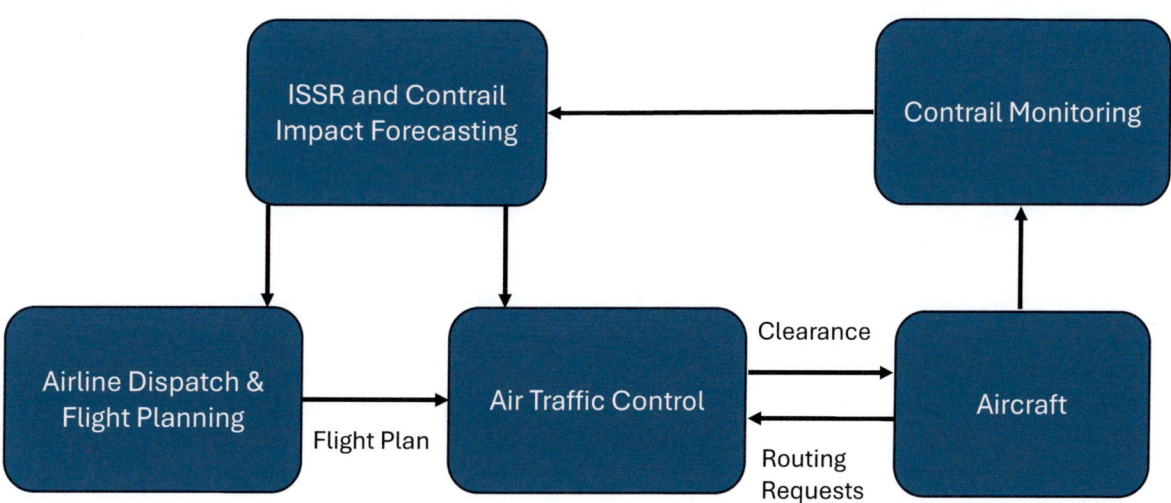

FIGURE 6-1 Information flow for a generic concept of operations for contrail avoidance.
NOTE: ANSP, Air Navigation Service Provider; ATC, air traffic control; ISSR, ice-supersaturated region; RF, radiative forcing.

The ability to implement operational contrail avoidance at scale will be limited by the capability of the ATC system to safely accommodate contrail-minimizing trajectories and altitudes. While only a small number of flights would need to avoid high-impact contrail-forming regions, there will be a natural concentration of traffic at the low-impact altitudes and the boundaries of the high-impact regions. The limitations of the current ATC system to accommodate optimal cruise altitudes can be seen in Figure 6-2, which shows the distribution of Automatic Dependent Surveillance-Broadcast (ADS-B) observed cruise altitudes for 1 year over the continental United States along with the corresponding distribution of cruise altitudes if each aircraft flew at the optimal altitude for minimum fuel burn. Aircraft commonly fly lower than the fuel-optimal cruise altitudes and must conform with standard flight levels, altitude for direction of flight conventions, and ATC restrictions (Figure 6-3).

Aircraft also often fly off of the optimal cruise altitudes when deviating for turbulence and ride quality reasons. This is normally managed by the local ATC controller on an individual flight basis and indicates that limited vertical maneuvering for contrail avoidance would be feasible in the near term.

Finding: Certain operational contrail mitigation concepts, focusing on forecasting large ISSRs with potential for warming contrails and moving all aircraft only vertically to minimize extra fuel burn (if possible), would more readily fit into the current ATC systems.

Because the current ATC system limits the ability of aircraft to fly at their optimal altitudes, advanced ATC approaches that increase vertical flexibility would have dual benefits, both for contrail avoidance and reduced fuel burn, thus reducing CO_2 emissions. NASA's Aeronautics Research Mission Directorate has a long history of developing future concepts and advanced ATC systems and would be well positioned to develop and investigate advanced ATC systems with the flexibility to safely accommodate routes avoiding high contrail-impact potential and reduced fuel burn and CO_2 emission as well as safety and turbulence avoidance.

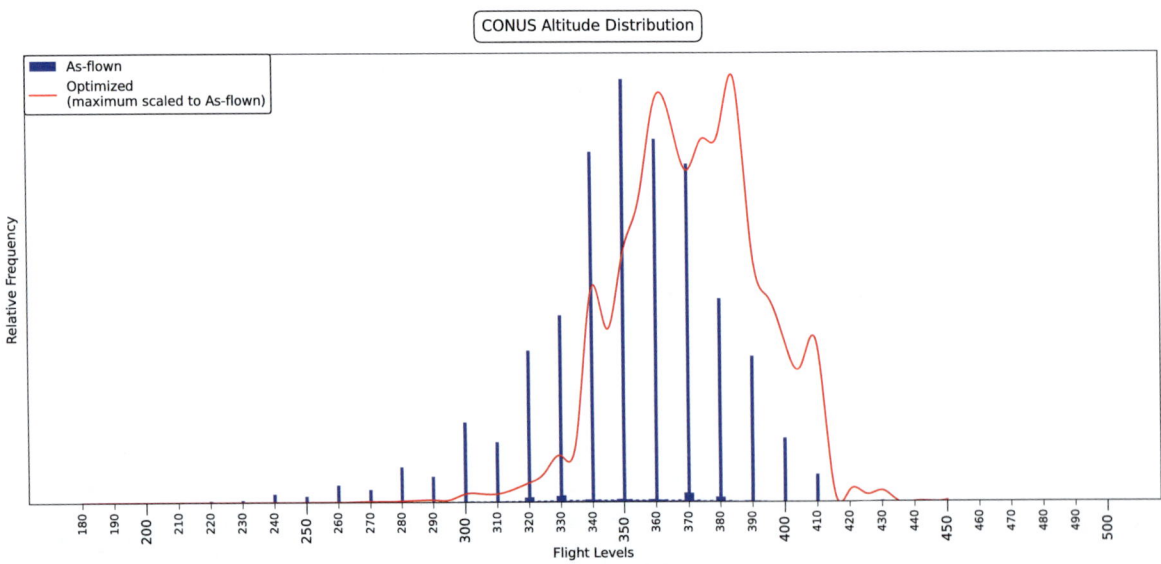

FIGURE 6-2 Observed distribution of commercial jet cruise altitudes (in blue) and fuel optimal cruise altitudes (in red) in the continental United States (CONUS) from November 2022 to October 2023. Distributions are normalized to have the same maximum value to allow comparison. Data include all flights for which aircraft performance data were available (excludes A321neo, 737MAX, CRJ, Dash 8).
SOURCE: Courtesy of Marek Travnik, Massachusetts Institute of Technology.

FIGURE 6-3 Contrails off the coast of Florida.

Finding: Contrail mitigation requires working within the current operational aviation system to safely reroute aircraft to avoid contrail-formation regions. Elements of the current system limit the ability of aircraft to do effective avoidance strategies.

Recommendation: NASA, in collaboration with airline operators and Air Navigation Service Providers, should continue research, development, and operational evaluation of advanced high-altitude air traffic control concept of operations to enable flexibility to accommodate fuel-efficient and contrail-avoidance flight trajectories.

REFERENCES AND FURTHER READING

Borella, A., O. Boucher, K.P. Shine, M. Stettler, K. Tanaka, R. Teoh, and N. Bellouin. 2024. "The Importance of an Informed Choice of CO_2-Equivalence Metrics for Contrail Avoidance." *Atmospheric Chemistry and Physics* 24:9401–9417. https://doi.org/10.5194/acp-24-9401-2024.

Engberg, Z., R. Teoh, T. Abbott, T. Dean, M.E.J. Stettler, and M.L. Shapiro. 2025. "Forecasting Contrail Climate Forcing for Flight Planning and Air Traffic Management Applications: The CocipGrid Model in Pycontrails 0.51.0." *Geoscientific Model Development* 18:253–286. https://doi.org/10.5194/gmd-18-253-2025.

Jensen, L., and R.J. Hansman. 2014. *Fuel Efficiency Benefits and Implementation Considerations for Cruise Altitude and Speed Optimization in the National Airspace System.* ICAT Report 2014-04. http://hdl.handle.net/1721.1/88517.

Kärcher, B. 2018. "Formation and Radiative Forcing of Contrail Cirrus." *Nature Communications* 9:1824.

Martin Frias, A., M. Shapiro, Z. Engberg, R. Zopp, M. Soler, and M.E.J. Stettler. 2024. "Feasibility of Contrail Avoidance in a Commercial Flight Planning System: An Operational Analysis." *Environmental Research: Infrastructure and Sustainability* 4(1):015013. https://doi.org/10.1088/2634-4505/ad310c.

A

Statement of Task

The National Academies of Sciences, Engineering, and Medicine will convene an ad hoc committee to develop a national research agenda to understand better, quantify, and support the development of technical and operational solutions to significantly reduce the global climate impact of aviation-induced cloudiness (AIC) and persistent contrails (PC) from commercial aviation. The research agenda will comprise a prioritized set of scientific recommendations and research projects critical to the national and international commercial aviation and scientific research communities. It will include a review of advances in measurements, modeling, technology, fuel sensitivities, and recent approaches for evaluating trade-offs.

The report will include recommendations for research to inform policymakers on appropriate metrics and methods to assess contrail-related climate impact and for research into technical and operational solutions to mitigate the impacts of AIC/PC. It will focus on understanding the importance of AIC/PC and the meaningfulness of various contributors to climate change so that federal agencies and industry can take further action to manage the important contributors. The prioritized research agenda developed by the committee will inform a contrail management strategy for the nation.

Specifically, the committee will

1. Consider the following:
   a. The current goals, guidance, and plans by government, industry, and other relevant entities to reduce global aviation CO_2 and non-CO_2 emissions.
   b. How the science of AIC/PC has changed over the last decade.
   c. The current state-of-the-art satellite imaging and modeling capabilities for identifying, characterizing, and predicting the formation of contrails and persistent contrail cirrus and how a change in altitude might reduce the climate effects.
   d. The role of sustainable aviation fuels (SAFs), alternative fuels (e.g., hydrogen), and state-of-the-art and future engine (propulsive power generation) technologies in impacting CO_2 and non-CO_2 emissions.
   e. The suite of aviation climate metrics and assumptions that includes the anthropogenic warming effects, such as CO_2 emissions and non-CO_2 impacts (nitrogen oxides and aviation-induced cloudiness).
2. Address the following:
   a. What capacity building is needed in terrestrial, airborne, and spaceborne data/imagery and modeling systems to improve the understanding and assessment of the climate effects of contrails, aviation-induced or modified cirrus clouds, and their climate impacts?

b. What capability may be needed for real-time relative humidity forecasting and measurements by aircraft in flight akin to wind, temperature, and speed, which are typically reported by aircraft? What are the advantages, disadvantages, and potential synergies between satellite and aircraft-based measurement?

c. What future research is needed to narrow the uncertainty and improve confidence in understanding the contrast between aviation-induced cloudiness and water vapor?

d. What future research would help characterize the impact of sustainable aviation fuels and other energy carriers on aviation-induced cloudiness?

3. Formulate a national research agenda to advance the scientific understanding of non-CO_2 emissions and AIC/PC impacts from commercial aviation, including but not limited to aviation fuels and propulsive power-generating technologies. The agenda should include

   a. The significant scientific, technical, economic, and policy challenges associated with this vision.

   b. A proposed research agenda consisting of a set of research projects, identified by priority groupings, that, if successful, could enable this vision.

   c. The agenda should be developed considering the resources and organizational partnerships required to complete the projects included in the plan.

   d. As appropriate, describe the potential contributions of U.S. research organizations, including NASA, other federal agencies (NOAA, FAA, DOE, etc.), industry, airlines, and academia. In addition, describe the operational roles of NOAA and FAA relevant to the formulation of a national research agenda. Finally, describe potential opportunities for international collaboration.

B

Committee and Staff Biographical Information

TIM LIEUWEN, *Chair*, is the executive vice president for research, a Regents' Professor, and holder of the David S. Lewis, Jr. Chair at the Georgia Institute of Technology. He is also the founder and chief technology officer of TurbineLogic, an analytics firm working in the energy industry. Dr. Lieuwen is an international authority on clean energy and propulsion and his work has contributed to numerous commercialized innovations in the energy and aerospace sectors. He has authored 4 books and more than 400 other publications. Current and past board positions include governing/advisory boards for Oak Ridge National Laboratory, Pacific Northwest National Laboratory, National Renewable Energy Laboratory, Electric Power Research Institute, and appointment by the Department of Energy (DOE) Secretary to the National Petroleum Counsel. He is an elected member of the National Academy of Engineering (NAE), a fellow of the American Society of Mechanical Engineers (ASME), the American Physical Society, and the American Institute of Aeronautics and Astronautics (AIAA) and a foreign fellow of the Indian National Academy of Engineering. Major awards include the ASME R. Tom Sawyer Award, AIAA Pendray Award, and ASME's George Westinghouse Gold Medal. Dr. Lieuwen has served on a variety of other National Academies of Sciences, Engineering, and Medicine studies, including Panel B: Propulsion and Power of the Decadal Survey on Civil Aeronautics; the Committee on Accelerating Decarbonization in the United States: Technology, Policy, and Societal Dimensions; the Committee on Advanced Technologies for Gas Turbines; and the Committee to Assess NASA's Flight Research Capabilities.

STEVEN BARRETT is the Regius Professor of Engineering at the University of Cambridge (United Kingdom). Prior to this, he was the H.N. Slater Professor and head of the Department of Aeronautics and Astronautics at the Massachusetts Institute of Technology (MIT) and the director of the MIT Laboratory for Aviation and the Environment. His research expertise is on advancing scientific understanding of aviation's environmental impacts and developing technical, operational, and fuel-based approaches to mitigating those impacts. One of his focus areas has been on modeling and observing contrails. This has included leading the development of advanced numerical models of contrail microphysical evolution and of machine learning approaches to detecting contrails in satellite imagery. He has also worked to develop and test operational contrail avoidance using near-real-time satellite observations. Other areas of research include characterizing aviation's emissions, evaluating the atmospheric chemistry impacts of those emissions, and assessing policy options for mitigating the environmental impacts of aviation. Barrett received all of his degrees in engineering from the University of Cambridge.

SEAN BRADSHAW is the senior technical fellow for sustainable propulsion at Pratt & Whitney, where his primary focus is the development of advanced propulsion technologies that will power the future of flight. Dr. Bradshaw is also engaged in Pratt & Whitney–sponsored university research and the development of aerospace engineering workforce capability. He provides leadership to the aviation industry through service as the chair of the ASME Committee on Sustainability, an associate editor of the ASME *Journal of Engineering for Gas Turbines and Power*, a member of the ASME Heat Transfer Committee, a member of the National Academies' Aeronautics and Space Engineering Board (ASEB) and Committee on Advanced Technologies for Gas Turbines, a former chair of the Gas Turbine Association, a former chair of the ASME Gas Turbine Technology Group, and an adjunct professor of mechanical engineering at Columbia University. Dr. Bradshaw earned a BS, an MS, and a PhD in aeronautics and astronautics from MIT.

LETICIA CUELLAR-HENGARTNER is a scientist at Los Alamos National Laboratory with more than 18 years of experience in stochastic and statistical modeling, risk assessment, and model validation of complex systems. She has served as the principal investigator (PI) on multiple high-profile initiatives, including an Ernst & Young–funded project developing forecasting models for audit risk, a study examining how human mobility affects disease propagation, and a co-led Probabilistic Effectiveness Methodology project assessing nuclear smuggling risks. Dr. Cuellar has authored or co-authored more than 100 technical reports and peer-reviewed journal articles, and she contributed to two National Academies' reports on integrating unmanned aircraft systems into the National Airspace System and evaluating the Transport Airplane Risk Assessment Methodology. She is currently a member of the National Academies' Community of Experts committee that advises the Federal Aviation Administration (FAA). Dr. Cuellar received a 2022 R&D 100 Award with special recognition as a Silver Medalist for Battling COVID-19 for her work on EpiGrid—an epidemiological software tool supporting decision making. She holds a master's degree and a PhD in statistics from the University of California, Berkeley.

ERIC H. DUCHARME retired in 2020 as the chief engineer at GE Aviation. He is currently the founder of an aerospace consultancy, Martlet Engineering, LLC. Dr. Ducharme has been responsible for leading many efforts and teams at GE, including composite aeroelastic technology and several jet engine engineering programs over their complete product life cycle. There he also led Advanced Technology Operation responsible for delivering technologies and architectures for next-generation commercial and military flight propulsion. Dr. Ducharme is an active member of the NAE. He is a fellow of AIAA, ASME, and Royal Aeronautical Engineering Society. He serves on the NAE/NASA Aeronautics Research and Technology Roundtable and the Aeronautics and Space Engineering Board. He is formerly a member of the NASA Aeronautics Committee.

ANDREW GETTELMAN is a senior scientist at Pacific Northwest National Laboratory specializing in cloud physics and global climate. He was a senior scientist at the National Center for Atmospheric Research helping lead climate model development efforts. Dr. Gettelman is an expert on the atmospheric and climate effects of aviation, contrails, and the physics and simulation of ice clouds, with more than 20 years of experience. He has written many papers on contrails, developed a global climate model with contrails, and participated in assessments and reviews of the effects of aviation on climate. Dr. Gettelman is a fellow of the American Geophysical Union (AGU) and the American Meteorological Society (AMS). He has been a Clarivate Highly Cited Researcher for the past 10 years. Dr. Gettelman is the author or co-author on more than 200 peer-reviewed publications and a textbook on climate modeling. He has a PhD in atmospheric sciences as well as a certificate in environmental management from the University of Washington. He received a BS in civil engineering from Princeton University. Dr. Gettelman previously served on the National Academies' Committee on Opportunities to Improve the Representation of Clouds and Aerosols in Climate Models with National Collection Systems.

ROBERT J. HANSMAN, JR., is the T. Wilson Professor of Aeronautics & Astronautics at MIT, where he is the director of the MIT International Center for Air Transportation. He conducts research in the application of information technology in operational aerospace systems. Dr. Hansman holds 7 patents and has authored more than 300 technical publications. He has more than 6,000 hours of pilot in-command time in airplanes, helicopters, and sailplanes including meteorological, production, and engineering flight test experience and received the Wright Brothers Master Pilot Award. Professor Hansman chairs the FAA Research Engineering and Development Advisory Committee. He is the co-director of the national Center of Excellence in Aviation Sustainability Center. He is a member of the NAE, a fellow of AIAA, and has received numerous awards, including the AIAA Dryden Lectureship in Aeronautics Research, the ATCA Kriske Air Traffic Award, a Laurel from Aviation Week & Space Technology, and the FAA Excellence in Aviation Award.

RICHARD H. MOORE is an airborne research scientist at the NASA Langley Research Center. He joined NASA in 2012 after receiving his PhD in chemical and biomolecular engineering from the Georgia Institute of Technology and BS and MS in chemical engineering from Bucknell University. His research examines the role of atmospheric aerosols in influencing cloud formation, air quality, and climate. He has participated in more than 20 airborne field campaign deployments in both instrument and project scientist roles and most recently served as the PI for the NASA Aeronautics–funded Boeing ecoDemonstrator Emissions Flight and Ground Tests. He designs and executes airborne field campaigns to measure aircraft engine emissions and contrails at cruise altitudes. A particular research interest is to understand how sustainable aviation fuels, low-particle-emitting engine technologies, and hydrogen-burning engines might impact the formation of climate-altering contrail cirrus clouds. Dr. Moore has been the recipient of the Presidential Early Career Award for Scientists and Engineers, a NASA Early Career Achievement Medal, and a NASA Aeronautics Research Mission Directorate Associate Administrator Award for Technology and Innovation. He is a member of AGU and AIAA and a recent member of the board of directors of the American Association for Aerosol Research.

JOYCE E. PENNER is the Ralph J. Cicerone Distinguished University Professor of Atmospheric Science in the Department of Climate and Space Sciences and Engineering at the University of Michigan. Dr. Penner's research focuses on improving climate models through the addition of interactive chemistry and the description of aerosols and their direct and indirect effects on the radiation balance in climate models. She is a fellow of AGU and AMS and was a winner of the Committee on Space Research William Nordberg Medal in 2022 and the AMS Syukuro Manabe Climate Research Award in 2021.

MICHAEL J. PRATHER, professor of Earth system science at the University of California, Irvine, retired but remains active in department governance and research. He was a Jefferson Science Fellow at the Department of State (2005) and a program manager at NASA Headquarters (prior to 1992). His studies of the chemistry and composition of the atmosphere focused on ozone depletion and climate change, authoring numerous World Meteorological Organization/United Nations Environment Programme and Intergovernmental Panel on Climate Change (IPCC) reports, including the 1999 IPCC Aviation Assessment. Dr. Prather's core research addresses the mathematical underpinnings of atmospheric chemistry and global biogeochemical cycles. He earned a BS (mathematics, Yale University), BA (physics, Oxford University), and PhD (astrophysics, Yale University). Dr. Prather has worked on numerous National Academies' studies that resulted in the following publications: *Causes and Effects of Stratospheric Ozone Reduction: An Update*, *Rethinking the Ozone Problem in Urban and Regional Air Pollution*, *A Climate Services Vision: First Steps Toward the Future*, *Weather Forecasting for FAA Traffic Flow Management: A Workshop Report*, *Global Sources of Local Pollution: An Assessment of Long-Range Transport of Key Air Pollutants to and from the United States*, *Verifying Greenhouse Gas Emissions: Methods to Support International Climate Agreements*, and *The Future of Atmospheric Chemistry Research: Remembering Yesterday, Understanding Today, Anticipating Tomorrow*.

STAFF

DWAYNE A. DAY is a senior program officer for the Aeronautics and Space Engineering Board (ASEB) of the National Academies. He has served as the study director for numerous studies for both ASEB and the Space Studies Board (SSB) on topics such as aeronautics research and technology, detecting and tracking potentially hazardous asteroids, planetary exploration, and astrophysics. Dr. Day has also published articles on space policy and history on subjects ranging from Cold War intelligence space programs to Apollo and the Space Shuttle. He served as an investigator on the Columbia Accident Investigation Board and previously worked for the Congressional Budget Office and the George Washington University Space Policy Institute. He received a PhD in political science from George Washington University.

LINDA M. WALKER is a program coordinator with ASEB, the Board on Physics and Astronomy (BPA), and SSB. Ms. Walker has been with the National Academies since September 2007. She has more than 45 years of administrative experience. Ms. Walker attended the University of the District of Columbia and Strayer University, majoring in business administration with a minor in human services.

DIONNA WISE is a program coordinator with SSB, having previously worked for the National Academies' Division of Behavioral and Social Sciences and Education. Recently, she was the lead study coordinator for the 2020 astronomy and astrophysics decadal survey. Wise has a long career in office administration, having worked as a supervisor in several capacities and fields. She attended the University of Colorado, Colorado Springs, majoring in psychology.

COLLEEN N. HARTMAN joined the National Academies in 2018, as the director for both SSB and ASEB, and became the director of the Board on Physics and Astronomy (BPA) in 2021. She is currently the senior director, aeronautics, astronomy, physics, and space science (SSB, BPA, ASEB). After beginning her government career as a presidential management intern under Ronald Reagan, Dr. Hartman worked on Capitol Hill for House Science and Technology Committee Chair Don Fuqua, as a senior engineer building spacecraft at NASA Goddard Space Flight Center, and as a senior policy analyst at the White House. She has served as the Planetary Division director, deputy associate administrator and acting associate administrator at NASA's Science Mission Directorate, deputy assistant administrator at the National Oceanic and Atmospheric Administration, and deputy center director and director of science and exploration at the NASA Goddard Space Flight Center. Dr. Hartman has built and launched scientific balloon payloads, overseen the development of hardware for a variety of Earth-observing spacecraft, and served as the NASA program manager for dozens of missions, the most successful of which was the Cosmic Background Explorer (COBE). Data from the COBE spacecraft gained two NASA-sponsored scientists the Nobel Prize in physics in 2006. She also played a pivotal role in developing innovative approaches to powering space probes destined for the solar system's farthest reaches. While at NASA Headquarters, she spearheaded the selection process for the New Horizons probe to Pluto. She helped gain administration and congressional approval for an entirely new class of funded missions that are competitively selected, called "New Frontiers," to explore the planets, asteroids, and comets in the solar system. She has several master's degrees and a PhD in physics. Dr. Hartman has received numerous awards, including two prestigious Presidential Rank Awards.

C

Recommendations

SHORT-TERM PRIORITY RECOMMENDATIONS

Recommendation: NASA should support the development, testing, and certification of advanced and accurate commercial-aircraft-capable humidity and temperature sensors for contrail-forming regions as well as onboard contrail-detecting cameras and automated contrail detection image recognition algorithms. (Chapter 3)

Recommendation: NASA should support research and observational studies to improve the understanding of the extent and frequency of ice-supersaturated regions (ISSRs) and the level of skill in simulating ISSRs and contrails. (Chapter 3)

Recommendation: NASA should apply its current Earth system modeling efforts in support of simulating ice-supersaturated regions and contrails as a pathway to demonstrate the use of observations and advanced modeling tools for developing a contrail forecast and prediction system and estimating contrail radiative forcing. (Chapter 5)

LONG-TERM PRIORITY RECOMMENDATIONS

Recommendation: NASA, in coordination with the Federal Aviation Administration, the Department of Energy, and the Department of Defense, should support laboratory and engine research studies to improve the understanding of how fuel composition, combustor technology, and engine operating conditions impact particulate emissions (volatile and non-volatile) and contrail properties. (Chapter 2)

Recommendation: NASA should continue to collect in-flight observational data of contrails, cruise emissions (CO_2, NOx, and ice-nucleating particles) from aviation that advance the understanding of the factors that influence contrail properties. (Chapter 2)

Recommendation: NASA should identify and enable a minimum set of key aerosol instruments that can be flown on multiple missions with the goal of characterizing the aerosol composition of the upper troposphere and uncovering the contribution of aviation emissions relative to other sources. (Chapter 3)

Recommendation: NASA, in coordination with the Federal Aviation Administration, the Department of Energy, the Department of Defense, other relevant federal agencies, and the private sector, should support development of low-particle-emitting combustion technologies, as well as sustainable aviation fuels with inherently low particulate-formation tendencies. (Chapter 2)

Recommendation: NASA should support observing system simulation experiments to define widespread water vapor sensor deployment to best inform contrail forecasts systems and individual verification and avoidance efforts. (Chapter 3)

Recommendation: NASA should support satellite remote sensing research for diagnosing persistent contrails and ice-supersaturated regions to develop readiness for the next-generation geostationary sounders and imagers. (Chapter 3)

Recommendation: As part of a national strategy, NASA should support development and assessment of models for all scales of contrail prediction. These models range from wake vortex to global climate to contrail plume to ice-supersaturation forecasting. (Chapter 4)

Recommendation: NASA should support development of a global contrail observing system as a foundation for research, analysis, and future verification. (Chapter 5)

Recommendation: NASA, in collaboration with airline operators and Air Navigation Service Providers, should continue research, development, and operational evaluation of advanced high-altitude air traffic control concepts of operations to enable flexibility to accommodate fuel efficient and contrail avoidance flight trajectories. (Chapter 6)